Physics: A First Course
Second Edition
Copyright © 2018 CPO Science, a member of School Specialty Science
ISBN: 978-1-62571-848-8
Part Number: 1576059
Printing 1—February 2017
Printed by Webcrafters, Inc., Madison, WI

All rights reserved. No part of this work may be reproduced or transmitted in any form or by any means, electronic or mechanical, including photocopying and recording, or by any information storage or retrieval system, without permission in writing. For permission and other rights under this copyright, please contact:

CPO Science
80 Northwest Boulevard
Nashua, NH 03063
(800) 932-5227
http://www.cposcience.com

Printed and Bound in the United States of America

Contributors

Principal Writers

Thomas C. Hsu, PH.D – Program Author
Nationally-recognized innovator in science and math education and the founder of CPO Science. Holds a Ph.D. in Applied Plasma Physics from the Massachusetts Institute of Technology (MIT), and has taught students from elementary, secondary, and college levels. Tom has worked with numerous K–12 teachers and administrators and is well known as a consultant, workshop leader, and developer of curriculum and equipment for inquiry-based learning in science and math.

Erik Benton – Investigation Writer
B.F.A., University of Massachusetts with minor in Physics
Taught for eight years in public and private schools, focusing on inquiry and experiential learning. Erik brings extensive teaching and technical expertise, ranging from elementary and adult education to wildlife research. Currently he is involved in bird population studies in Massachusetts. Erik is the CPO Science investigation writer and conducts national content presentations.

Scott Eddleman – Project Manager and Writer
B.S., Biology, Southern Illinois University; M.Ed., Harvard University
Taught for 13 years in urban and rural settings. Developed two successful science-based school-to-career programs. Participated in a fellowship at Brown University where he conducted research on the coral reefs of Belize. Worked on National Science Foundation–funded projects at TERC. Scott has been a principal writer and curriculum developer for CPO Science since 1999.

Patsy Eldridge – Principal Writer
B.S., Biology, Grove City College; M.Ed., Tufts University
Experienced science teacher and national hands-on science trainer and presenter. Develops and teaches content-intensive graduate courses for science educators. Worked as a research scientist in the medical device industry. Patsy has developed curriculum and training materials with CPO Science since 2000.

Laine Ives – Writer
B.A., Gordon College and graduate coursework at Cornell University's Shoals Marine Laboratory and Wheelock College
Taught elementary and middle school science at an independent school and an environmental education center in New England. Taught middle and high school English overseas. Laine has worked in curriculum development at CPO Science since 2000.

Stacy Kissel – Writer
B.S., Civil and Environmental Engineering, Carnegie Mellon University; M.Ed., Physics Education, Boston College
Teaches physics, math, and integrated science at Brookline High in Massachusetts. Stacy was selected as Brookline's teacher of the year for 2007.

Sonja Taylor – Principal Teacher's Guide Writer
B.S., Chemistry, Stephen F. Austin State University; M.Ed., Divergent Learning, Columbia College
Worked as an analytical chemist and has taught high school chemistry, biology, and physical science. While teaching in an innovative academy-based high school, Sonja was an academy team leader with responsibilities that included overseeing cross-curricular integration and serving as the liaison between students, faculty and members of the corporate advisory board. She enjoys developing lessons focused on engaging learners at all levels. Sonja has been a writer for CPO Science since 2003.

Senior Editor

Lynda Pennell – Executive Vice President
B.A., English; M.Ed., Administration, Reading Disabilities, Northeastern University; CAGS Media, University of Massachusetts, Boston.
Nationally known in high school restructuring and for integrating academic and career education. Served as the director of an urban school for 5 years and has 17 years teaching/administrative experience in the Boston Public Schools. Lynda has led the development for CPO Science for the past eight years. She has also been recognized for her media production work.

Editorial Consultant

Alan Hull
B.S., Emery University, Atlanta, Georgia; M.S., Georgia State University
Former high school physics and math teacher in Atlanta, Georgia. After teaching, Alan entered the school textbook publishing business where he held a variety of positions, including as a national secondary math-science consultant for major publishing companies.

Art and Illustration

Polly Crisman – Graphics Manager/Illustration
B.F.A., University of New Hampshire
Worked as a designer and illustrator in marketing and advertising departments for a variety of industries. Polly has worked in the CPO Science design department since 2001, and is responsible for organizing workflow of graphics and file management. She created the CPO Science logo and supervises the graphic design image for CPO publications and media products.

Jesse Van Valkenburgh – Illustration/Photography
B.F.A. Illustration, Rochester Institute of Technology
Worked in prepress and design. Was responsible for creative design and prepress film production for computer catalogs, brochures, and various marketing materials. Jesse completes photography and illustrations as a graphic designer for CPO Science book and media products.

Bruce Holloway – Senior Designer/Illustrator
Pratt Institute, New York, Boston Museum of Fine Arts
Expertise in illustration, advertising graphics, exhibits and product design. Commissioned throughout his career by The National Wildlife Federation's Conservation Stamp Campaign. Other commissions include the New Hampshire State Duck Stamp campaigns for 1999 and 2003.

Equipment Design and Material Support

Thomas Narro – Senior Vice President
B.S., Mechanical Engineering, Rensselaer Polytechnic Institute
Accomplished design and manufacturing engineer; experienced consultant in corporate reengineering and industrial-environmental acoustics.

Marise Evans – Industrial Designer
M.S., Industrial Design, Auburn University, B.S., Interior Design and Environmental Design, Auburn University
Works with mechanical engineers to design new CPO products and improve the design of current products.

Dr. Darren Garnier – Research Physicist, Columbia University/MIT

Project and Technical Support

Susan Gioia – Educational CPO Science Administrator
Expertise in office management. Oversees all functions necessary for the smooth product development of CPO products, including print and media.

Sara Desharnais – Electronic Production Specialist
B.A., Creative Writing, Chester College of New England
Has worked in the publishing industry since 2004 and joined CPO Science in 2010.

Assessment

Mary Ann Erickson
B.S., Naval Architecture and Marine Engineering, Massachusetts Institute of Technology
Ran a technical writing consulting business, writing process control manuals for water treatment plants, software design documentation for simulation software, and operator manuals for mining equipment.

David H. Bliss
B.S., Science, Cornell University; M.Ed., Zoology
Taught for 39 years in the science field: biology, chemistry, earth science, and physics. Served as science department chair of Mohawk Central School District in Mohawk, New York.

Jane Fisher
B.S. Economics, Massachusetts Institute of Technology; Associate of the Society of Actuaries
Taught science for 39 years.

Materials Support

Kathryn Gavin – Purchasing and Quality Control Manager
Responsible for all functions related to purchasing raw materials and quality control of finished goods. Works closely with product development and design.

Matthew Connor – Product Quality Specialist
B.A., Philosophy, University of Toronto.
Responsible for production quality assurance, product testing and troubleshooting, creating and editing product documents and experiment testing.

Science Content Consultants

Dr. Jeffrey Williams – Bridgewater State College, Bridgewater, Massachusetts - Professor of Physics

Dr. David Guerra – Associate Professor, Department Chair Physics, St. Anselm College, Manchester, New Hampshire

Dr. Mitch Crosswait – Nuclear Engineer/Physicist, United States Government, Alexandria, Virginia

Science Content Reviewers

Dr. Jeff Schechter – Physicist, Boston, Massachusetts

Dr. Tim Daponte – Physics Teacher, Houston ISD, Houston, Texas

Wanda Pagonis – Physics Teacher, Lady of the Lake University, San Antonio, Texas

Kurt Lichtenwald – Physics/Robotics Teacher, Gloucester High School, Gloucester, MA

Dr. Willa Ramsay – Science Education Consultant

DaMarcus Wright – Physics Teacher, Grand Prairie ISD, Grand Prairie, Texas

Beverly T. Cannon – Physics Teacher, Highland Park High School, Dallas, Texas

Betsy Nahas – Physics Teacher, Chelmsford High School, Westford, Massachusetts

Neri Giovanni – Physics Teacher, Kennedy High School, San Antonio, Texas

Scott Hanes – Physics Teacher, Liberty-Eylau High School, Hooks, Texas

Bruce Ward – Nuclear Medical Technician, Boston, Massachusetts

Lee DeWitt – Physics Teacher, NEISD, San Antonio, Texas

Steve Heady – Physics Teacher, Houston High School, Houston, Texas

Dr. Michael Saulnier – Physicist, Boston, Massachusetts

Gigi Nevils – Physics Teacher, Bellaire High School, Houston, Texas

Lebee Meehan – Physicist, National Aeronautics and Space Administration, Houston, Texas

Dr. Manos Chaniotakis – Physicist, Massachusetts Institute of Technology, Cambridge, Massachusetts

Jay Kurima – Physics Teacher, Fort Worth ISD, Fort Worth, Texas

James DeHart – Physics Teacher, West Brook High School, Beaumont, Texas

David Binette – Engineering Student, Cornell University, Ithaca, New York

Rebecca DeLozier – Physics Teacher, Lewisville ISD, Shady Shores, Texas

Valerie Felger – Physics Teacher, North East ISD, New Braunfels, Texas

George Whittemore – Physics Instructor, Leominster High School, Leominster, MA

Ken Rideout – Physical Science Teacher, Swampscott High School, Swampscott, MA

Brett Malas – National Board Certified Science Teacher, Naperville North High School, IL

William G. Fleischmann – Science Teacher, Wood Hill Middle School, MA

Ruby Ashley – Project Coordinator, Georgia Southern Museum Outreach Programs, GA

Mary Jo Carabatsos PhD – Science Teacher, Andover High School, Andover, MA

Angela Benjamin – AP Physics Instructor, Woodrow Wilson Senior High School, DC

Richard Famiglietti – Science Teacher, Lynn Classical High School, Lynn, MA

Elizabeth A. Jensen – Science Teacher, James Blair Middle School, VA

Colleen M O'Shell – Chemistry Teacher, Cambridge Rindge and Latin School, MA

Ian Smith – Physics Teacher, Bellows Free Academy, VT

Jean Lifford – Reading Coach, Boston Public Schools, MA

Ed Wiser – Physics Teacher, Brookline High School, MA

David Harris – Head of Science Department, Hackley School, NY

Sarah Segreti – Science Teacher, Naperville North High School, IL

Nick Nicastro – Physics Teacher, Wachusett Regional High School, MA

Cecilia A. Cherill – Physical Science Teacher, Churchill Junior High School, NJ

Kristy Beauvais – Physics Teacher, Concord-Carlisle Regional High School, MA

How to Read an Investigation

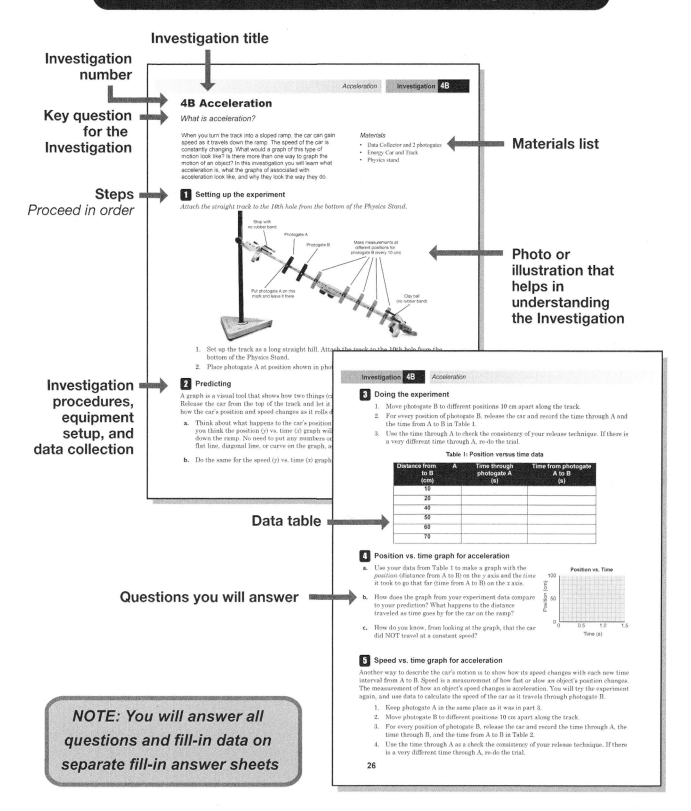

- Investigation number
- Investigation title
- Key question for the Investigation
- Steps — *Proceed in order*
- Materials list
- Photo or illustration that helps in understanding the Investigation
- Investigation procedures, equipment setup, and data collection
- Data table
- Questions you will answer

NOTE: You will answer all questions and fill-in data on separate fill-in answer sheets

LAB SAFETY

Observing safety precautions is an extremely important practice while completing science investigations. Using science equipment and carrying out laboratory procedures always requires attention to safety. The purpose of learning and discussing safety in the lab is to help you learn how to protect yourself and others at all times.

The investigations in this book are designed to reduce safety concerns in the laboratory. The physics investigations use stable equipment that is easy to operate. The chemistry investigations use both household and laboratory chemicals. Although these chemicals might be familiar to you, they still must be used safely.

You will be introduced to safety by completing a skill sheet to help you observe the safety aids and important information in your science laboratory. In addition to this skill sheet, you may be asked to check your safety understanding and complete a safety contract. Your teacher will decide what is appropriate for your class.

Throughout this book, safety icons and words and phrases like "caution" and "safety tip" are used to highlight important safety information. Read the description for each icon carefully and look out for them when reading your book and doing investigations.

Icon	Description
	General safety: Follow all instructions carefully to avoid injury to yourself or others.
	Wear safety goggles: Requires you to wear eye protection to prevent eye injuries.
	Wear a lab apron or coat: Requires you to wear a lab apron or coat to prevent damage to clothing and to protect from possible spills.
	Wear gloves: Requires you to protect your hands from injury due to heat or chemicals.
	Poisonous chemicals: Requires you to use extreme caution when working with chemicals in the laboratory and to follow all safety and disposal instructions from your teacher.
	Skin irritant: Requires you to use extreme caution when handling chemicals in the laboratory due to possible skin irritation and to follow all safety and disposal instructions from your teacher.
	Respiratory irritant: Requires you to perform the experiment under a laboratory hood and to avoid inhaling fumes while handling the chemicals.
	Laser: Requires you to use extreme caution while using a laser during investigations and to follow all safety instructions.

Lab safety is the responsibility of everyone! Help create a safe environment in your science lab by following the safety guidelines from your teacher as well as the guidelines discussed in this document.

Investigations

1.1	Measuring Time	1
1.2	Speed	3
1.3	Experiments and Variables	8
2.1	Position, Speed, and Velocity	10
2.2	Position, Velocity, and Time Graphs	13
2.3	Accelerated Motion	15
3.1	The Law of Inertia	18
3.2	The Second Law: Force, Mass, and Acceleration	20
3.3	Free Fall	23
4.1	Newton's Third Law and Momentum	25
4.2	Conservation of Energy	27
4.3	Collisions	31
5.1	Working with Force Vectors	34
5.2	Equilibrium of Forces and Hooke's Law	38
5.3	Friction	42
6.1	Launch Angle and Range	44
6.2	Launch Speed and Range	48
6.3	Levers and Rotational Equilibrium	50
7.1	Force, Work, and Machines	54
7.2	Work and Energy	58
7.3	Energy and Efficiency	61
8.1	Energy Flow in a System	63
8.2	People Power	65
8.3	Transportation Efficiency	67
9.1	Temperature and Heat	69
9.2	Energy and Phase Changes	71
9.3	Exploring Heat Transfer	73
10.1	The Atom	77
10.2	Energy and the Quantum Theory	80
10.3	Nuclear Reactions and Radioactivity	82
11.1	Frames of Reference	84
11.2	Special Relativity	87
11.3	General Relativity	90
12.1	Measuring Voltage and Current	91
12.2	Resistance and Ohm's Law	95
12.3	Building an Electric Circuit Game	98
13.1	Series Circuits	100
13.2	Parallel Circuits	103

13.3	Electrical Energy and Power	105
14.1	Electric Charge	108
14.2	The Flow of Electric Charge	110
14.3	Making an Electrophorus	113
15.1	Magnetism	115
15.2	Electromagnets	117
15.3	Making a Model Maglev Train	120
16.1	Electromagnetic Forces	122
16.2	Electromagnetic Induction	125
16.3	Generators and Transformers	128
17.1	The Magnetic Field	131
17.2	Using Fields	133
17.3	Electric Forces and Fields	135
18.1	Harmonic Motion and the Pendulum	138
18.2	Harmonic Motion Graphs	141
18.3	Natural Frequency	144
19.1	Waves in Motion	147
19.2	Resonance and Standing Waves	150
19.3	Exploring Standing Wave Properties	152
20.1	Sound and Hearing	154
20.2	Properties of Sound Waves	156
20.3	Sound as a Wave	158
21.1	Properties of Light	163
21.2	Additive Color Model and Vision	165
21.3	Subtractive Color Model	168
22.1	Reflection	169
22.2	Refraction	171
22.3	Images from Mirrors and Lenses	174
23.1	The Electromagnetic Spectrum	180
23.2	The Wave Nature of Light—Polarization	181
23.3	The Particle Nature of Light—Phosphorescence	184

Extension Investigations

Engineering Design Log	187
Chapter 3: Newton's Second Law	198
Chapter 4: Momentum and the Third Law	204
Chapter 6: Engineering Design and Projectile Motion	209
Chapter 10: Nuclear Reactions Game	214

1.1 Measuring Time

How do we measure and describe time?

In science, it is often important to know how things change with time. The DataCollector allows us to make accurate, precise time measurements by performing many different functions. You will explore several of these functions, some with the use of a photogate.

Materials List
- DataCollector
- Photogate

In this investigation, you will:

- time events in stopwatch mode and with a photogate.
- consider how accuracy, precision, and resolution are important to time measurements.

1 Stopwatch mode

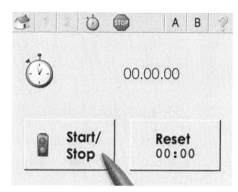

In stopwatch mode, the DataCollector is used to measure time intervals. The stopwatch icon indicates the DataCollector is in Stopwatch mode. The clock is started and stopped by tapping Start/Stop. To reset the clock, tap Reset. The display shows time in seconds, and then it changes to show minutes and seconds for times longer than one minute.

1. At the DataCollector home screen, choose stopwatch mode.
2. Practice starting, stopping, and resetting the stopwatch.
3. Start and stop the stopwatch as quickly as you can. Who in your group can get the shortest time? Each group member should perform several trials. Find the average time for each group member.

a. Does this stopwatch measure time to the nearest tenth, hundredth, or thousandth of a second?

b. What was your average start/stop time? What is the average start/stop time for the entire class?

c. In general, what does your start/stop time represent? (*Hint*: What would you call the time it takes your body to send a signal from your brain to a muscle?)

d. Does everyone in the class have the same average start/stop time? Why or why not?

e. What variables could affect start/stop time?

f. When you start and stop the stopwatch repeatedly, do you always get the same time? Why or why not?

g. Could this stopwatch be used to time a marble falling from a table down to the floor? Why or why not?

2 Using the photogate

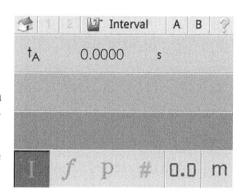

1. Connect a single photogate to input A with a cord.
2. From the DataCollector home screen, choose CPO Timer mode. There are four different functions for this mode, which are displayed at the bottom of the screen. Choose Interval by tapping the interval function button "I." A photogate icon at the top of the screen indicates a photogate is connected to the DataCollector. The word *Interval* next to it indicates the Interval function of Timer mode has been selected.

Investigation 1.1 Measuring Time

3. The photogate uses a light beam to start and stop the DataCollector. Try blocking the light beam with your finger and observe what happens to the t_A value on the screen.

4. Which group member can get the shortest time for t_A? Use your finger, a pencil, or some other object, as long as you don't do anything that would damage the photogate.

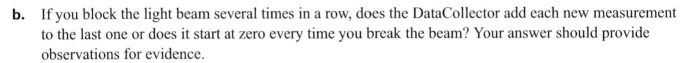

a. What action do you take to start and stop the DataCollector? Be very specific in your answer. Someone who has never seen the photogate before should be able to read your answer and know what to do with the light beam to make the DataCollector start and stop.

b. If you block the light beam several times in a row, does the DataCollector add each new measurement to the last one or does it start at zero every time you break the beam? Your answer should provide observations for evidence.

c. Does t_A display time to the nearest tenth of a second, hundredth of a second, thousandth of a second, or some other fraction of a second? Explain.

d. What was the shortest time for t_A in your group? How does this time compare to the shortest stopwatch time your group achieved?

e. Suppose you are timing how long it takes a marble to fall from a table to the floor. Could you use the DataCollector with the photogate to time this? Why or why not?

3 Accuracy, resolution, and precision

a. *Resolution* refers to the smallest interval that can be measured. Use a photogate to determine the resolution of the DataCollector. Give your answer in seconds and tell how your observations support your answer.

Accurate — Accurate and precise

b. The word *accuracy* refers to how close a measurement is to the true value. Which of the following statements best describes what you know about the accuracy of time measurements made with a photogate? Give a reason for your answer.

1. The DataCollector is accurate to 0.001 seconds.
2. The DataCollector is accurate to 0.0001 seconds.
3. It is impossible to know the accuracy without more information about how the DataCollector determines 1 second.
4. A time of 0.0231 seconds is more accurate than a time of 26 seconds.

Precise but not accurate — Not precise and not accurate

Low resolution — High resolution

c. The word *precision* describes how closely repeated measurements of *the same quantity* can be made. When multiple measurements are very precise, they are close to the same value. For example, an ordinary clock (with hands) can determine the time to a precision of about a second. That means many people reading the same clock at the same time will read times that are within a second of each other. However, it is possible to be precise but not accurate. Which is likely to be more precise: time measurements made with a stopwatch or measurements made with photogates? Explain.

1.2 Speed

Can you predict the speed of the car as it moves down the track?

What happens to the speed of a car as it rolls down a ramp? Does the speed stay constant or does it change? In this investigation, you will measure the speed of a car at different points as it rolls down a ramp. Then you will make a graph that describes the motion, and predict the speed of the car somewhere on the ramp.

Materials List
- DataCollector
- Photogate
- Energy Car
- SmartTrack
- Physics Stand
- Metric ruler
- Graph paper

1 Describing speed

Suppose you ran in a race. What information do you need to describe your speed? Saying that you ran for 20 minutes would not be enough information. To describe your speed, you need to know two things:

1. the **distance** you traveled, and
2. the **time** it took you to travel that distance.

Example	Distance	Time	Speed
10 seconds / 100 meters (cheetah)	100 meters	10 seconds	10 m/s
1 hour / 50 miles (car)	50 miles	1 hour	50 mph
15 seconds / 10 feet (fish)	10 feet	15 seconds	0.67 ft/s

Based on the examples above, fill in the boxes to complete the *equation* for calculating speed.

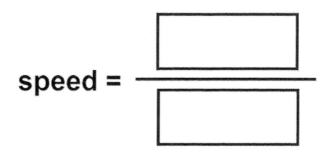

Investigation   1.2   Speed

2 Making a hypothesis

Where will the car be going the fastest on the ramp: At the top, middle, or bottom?

a. Create a hypothesis about the car's speed on the ramp—where will it be the fastest, and why do you think so?

3 Setting up the experiment

1. Attach the SmartTrack to the physics stand using one threaded knob. Your teacher will tell your group which hole on the stand to use. Hole where ramp is attached: _____

2. Connect the cord to the photogate and the A input on the top of the DataCollector.

3. Place the photogate on the SmartTrack.

4. Turn on your DataCollector and select Timer mode from the home screen.

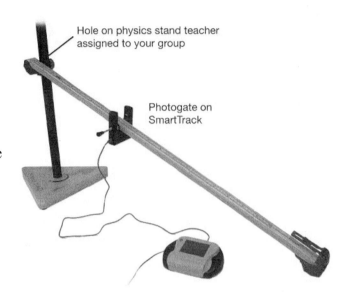

4 Using the photogate to measure speed

As the car passes through the photogate, the DataCollector's clock starts and stops. The DataCollector measures the length of time that the light beam of the photogate is broken. Speed is equal to the distance traveled divided by the time taken. What distance should you use?

If you look at the car you will see a small "flag" on one side. This is the part of the car that blocks the photogate's light beam.

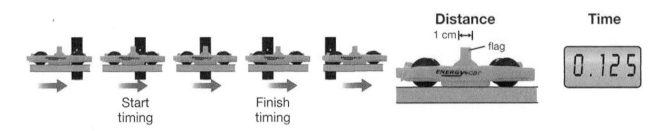

a. What is the *distance* traveled by the car in the photo?

b. Put the car onto the SmartTrack. Notice how the photogate's light is green. What happens to the green light when the photogate's beam is blocked? How can you tell it is no longer being blocked?

c. Use a ruler to measure how far the car rolls while the flag on the car blocks the beam. How far does the car travel while breaking the photogate's beam?

d. What distance should you use to calculate the car's speed when moving through photogate A?

e. What is the *time* taken by the car in the photo?

f. What is the *speed* of the car in the photo?

5 Doing the experiment

1. Place the car at the top of the ramp and hold it in place by pressing the hold button and resting the front of the car on the holding pin that comes up through the track.

2. Measure 10 centimeters from the front edge of the car's flag down the track. Place the photogate so its beam is exactly 10 centimeters down the hill from the wing of the car. You can tell where the beam is located by finding the crosshairs on the inside and outside of the photogate. Record the photogate's position as 10 cm in Table 1.

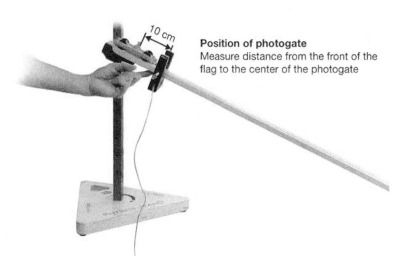

Position of photogate
Measure distance from the front of the flag to the center of the photogate

3. Release the hold button and allow the car to roll down the hill. Record the time through photogate A at the 10 cm position in Table 1.
4. Calculate the speed of the car using the distance traveled (1 cm) and the time at photogate A.
5. Move the photogate at least 5 centimeters further down the track. Record the position, time, and speed in the second row of Table 1.
6. Repeat measurements of position, time, and speed for at least four more different places on the ramp.

Table 1: Position, time, and speed data

Position of photogate A (cm)	Time through photogate A (s)	Distance traveled by the car (cm)	Speed of the car (cm/s)
		1.00	
		1.00	
		1.00	
		1.00	
		1.00	
		1.00	

5

Investigation 1.2 Speed

6 Analyzing the data

a. From your measurements, what can you tell about the car's speed as it moves down the ramp?

b. Use your data to make a graph which shows how the car's speed changes as it rolls down the ramp. Put speed on the *y*-axis and position of the photogate on the *x*-axis. Be certain to label the axes with the correct variable and the proper unit of measurement. Give the graph a descriptive title. Include the number of the hole you used to connect the ramp to the stand in your title.

c. Describe what the graph shows about how the speed of the car is changing as it moves down the ramp.

d. Compare your graph with that of students who connected their ramps at different heights on the stand. Explain any differences you see.

7 Making a prediction

Now that you have gathered, organized, and analyzed your data, it is time to make a prediction. You measured the speed of the car at several places on the ramp as it rolled to the bottom. Now, you will predict what the speed of the car would be at a place you did not measure. There is a way to do this with the information represented by your graph.

1. In Table 2, record a position on the ramp at which you did not measure the speed of the car. The position should be between two places where you did measure the speed.

Table 2: Predicted speed data

Selected position (cm)	Predicted speed at selected position (cm/s)	Actual speed at selected position (cm/s)	Percent correct of prediction

2. Use your graph to find the predicted speed of the car at the selected position. To do this, start on the *x*-axis at the position you have selected. Draw a line straight up until it intersects the speed vs. position line on your graph. At the intersection point, draw a line horizontally over to the *y*-axis where the speed is recorded. This is the speed that corresponds to your predicted location. The graph to the right uses a position of 55 cm as an example. (You should use a different position.) Record your predicted speed in Table 2.

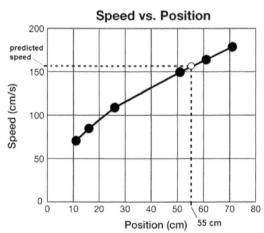

3. Place the photogate at the position you selected in step 1 and record the time it takes for the car to pass through the photogate.

4. Use the wing length (1.00 cm) and the time to calculate the speed. Record the actual speed in Table 2.

5. How does your predicted speed compare with the actual measured speed? What does this tell you about your investigation and measurements?

8 Analyzing data for error

a. Find the difference between the predicted speed and the actual, calculated speed.

   Predicted speed − Actual speed = Difference

b. Take this difference and divide it by the actual speed, then multiply by 100 to get the percent error.

   (Difference ÷ Actual speed) × 100 = Percent error

c. Use the percent error to calculate percent correct. Record percent correct in Table 2.

   100 − Percent error = Percent correct

d. What do you think can account for any error in your prediction?

1.3 Experiments and Variables

How do you plan an investigation?

Investigations help us collect evidence so we can unlock nature's puzzles. If an investigation is well planned, the results can provide an answer to a scientific question like "What would happen if I did this?" If the investigation is not well planned, you will still get results, but you may not know what they mean. In this investigation, you will experiment with a car on a ramp. Only by paying careful attention to the variables can you make sense of the results.

Materials List
- Energy Car
- SmartTrack
- DataCollector
- 2 Photogates

1 Setting up the experiment

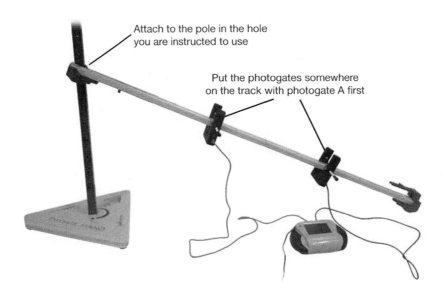

1. Attach the SmartTrack to the physics stand with a blue threaded knob. Your teacher will tell you which hole in the stand you should use. Each group's track will have a different angle.
2. Place two photogates on the SmartTrack and connect them both to the DataCollector. Be sure to place photogate A higher up the hill than photogate B.
3. Roll the car down and use Table 1 to record the time it takes the car to pass between the photogates (t_{AB}).

2 Making a prediction

a. Which track should have the fastest car? Which track should have the shortest time between photogates?

b. Write a one-sentence hypothesis that relates the time between photogates to the angle of the track.

c. Use Table 1 to record the results from each group in your class. Record the times in the column labeled "First trial." Leave the column labeled "Second trial" blank. How do the results compare with your hypothesis? Can you give a reason why they did or did not behave as you expected?

Table 1: Photogate times from A to B

	First trial	Second trial
Attachment hole (holes from bottom)	Time from A to B (s)	Time from A to B (s)

3 Variables

a. List at least six variables in your system that affect the time between photogates.

b. Which variable is the experimental variable in your class? How do you know?

c. What should be done with the other variables (other than the experimental variable)? Why should this be done?

d. Name two variables that should not be included in your system. These variables should not have much (or any) influence on the time from photogate A to B.

4 A controlled experiment

1. With your teacher and the rest of your class, decide how to control the variables other than the experimental variable.
2. Practice rolling the car until you can get three consecutive trials within 0.0010 seconds of each other.
3. Repeat the experiment using the experimental and controlled variables you discussed and decided upon. Record the new data in the column titled "Second trial."

5 Arguing from evidence

a. Did the second trial of the experiment produce results that agree with your hypothesis?

b. Why does the second trial produce better agreement with your hypothesis than the first trial did?

c. If something does not work, discuss what you should do to try and find the problem. List at least three steps that relate to variables, experiments, and controls.

2.1 Position, Speed, and Velocity

How are position, speed, and velocity related?

Knowing an object's position and measuring how fast or slow it is moving can be helpful, but not always precise enough for science. Sometimes it is also important to know in what direction an object is moving.

In this investigation, you will use a model to:
- measure increases and decreases in position values.
- measure positive and negative velocity.
- compare speed and velocity.

Materials List
- DataCollector
- SmartTrack
- Energy Car
- Graph paper
- Bubble level
- Velocity sensor

1 Position

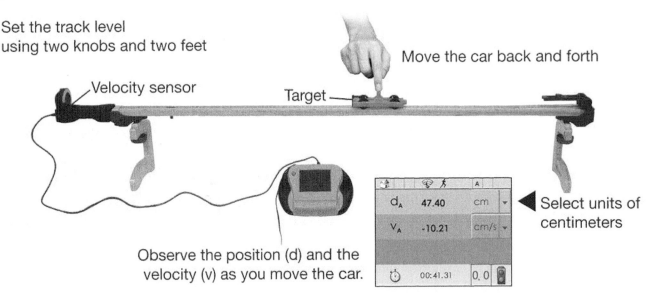

1. Attach the velocity sensor to the track, then attach the communication cable so that one end is plugged into the sensor and the other is plugged into the A slot of the DataCollector.
2. Set up the SmartTrack so it is approximately level using the bubble level. You may need to slide a few sheets of paper under one end of the track to make it level.
3. Set the DataCollector to Meter mode. The units will default to centimeters for position, or distance, (d) and centimeters/second for velocity (v).
4. Attach the target to the car and place the car at the center of the track so the target is positioned near the 50-cm mark, initially leaving the car at rest.
5. Move the car backward (toward the sensor) and forward (away from the sensor) along the track. Observe the values of position and velocity. Answer the questions below using this setup.

2 Using your model

a. Did you observe both increasing and decreasing positions? If so, describe when the position increased and when it decreased.

b. Did your position value change sign? What do you think it would take to see a negative position value?

c. The default zero position is at the zero-centimeter mark on the SmartTrack. If the zero were at the center of the track, would you expect to see a sign change when you run this same experiment?

d. What is *velocity*? What determines when the velocity value changes sign?

3 Speed and velocity

1. Use the same setup of the EnergyCar and SmartTrack that you used for Part 1 of this investigation.
2. Start a new experiment by going to the Home screen and tapping Data Collection.

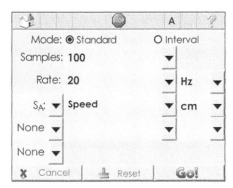

3. Tap Setup on the menu bar at the bottom of the screen. Set the samples to 100 and the rate to 20 Hz. This will give you five seconds to move the car back and forth on the track.

4. Select Speed and cm/s as your measurement input. The other two inputs should be set to None.

5. Once again, you will begin the experiment with the car at the center of the track, but instead of moving the car back and forth randomly, create a plan with your group for how you are going to move the car for the entire five seconds. Decide in which direction and how fast or slow you will move the car. Make at least three changes in direction, come to a complete stop at least once, and move the car fast at least once and slow at least once. Write out your plan using words and arrows indicating direction and speed. Use a step-by-step approach and describe what should be done during each step. Each step should describe some kind of change.

6. Hit the Go button to start the experiment.
7. In order to watch the velocity change in real time on a graph when you move your car on the track, switch to the Graph tab.
8. Move your car according to your plan and watch the graph it makes.
9. Take careful note of what the graph does as you move your car. Write these notes next to each instruction on your list. You will need to be able to make a connection between what is being displayed on the graph and how the car is moving.
10. To repeat the experiment with the current setup, tap the Setup tab and hit Repeat.

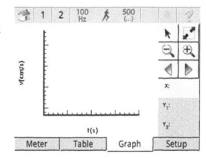

11

Investigation 2.1 Position, Speed, and Velocity

4 Arguing from evidence

a. Describe the graph you see. Make a careful sketch of the graph on a piece of graph paper and label the axes.

b. When is the velocity positive? When is it negative?

c. Velocity is a change in position divided by a change in time. Speed is a change in distance divided by a change in time. What is the difference between a change in position and a change in distance?

d. Look at your sketch of the velocity vs. time graph. If you were to use this graph to make a speed vs. time graph, how would the two graphs be different? How would the two graphs be similar?

e. Use your sketch of the velocity vs. time graph to sketch a speed vs. time graph. Do you think you could make an accurate velocity vs. time graph from a speed vs. time graph? Why or why not?

f. Look at your sketch of your velocity vs. time graph. Match up the steps from your plan with the places on the graph where they occur. Describe what each step created on your graph. Be descriptive, and state what the shape of the graph looks like at each step: if it created a straight line or a curve, whether it is sloping up or down, and whether it is in the positive or negative region of your graph.

g. Set up the experiment as you did for section 3. Instead of written steps, use the sketch you made from Part 3 as your guide, and try to match the DataCollector's graph to the sketch as you move the car. Describe how well your group was able to match the graph, what you found difficult, and what you found easy. Keep trying until you get a good match.

h. *Challenge*: Set up the experiment again. Trade sketches with another group. See if you can match their sketch with your DataCollector's graph as you move the car using their sketch as your guide. Trade sketches with as many groups as time permits. Describe your successes and challenges as you match their graphs. Make a speed vs. time sketch based on each velocity vs. time sketch you try to match.

2.2 Position, Velocity, and Time Graphs

How are graphs used to describe motion?

A graph presents information in a format that can be understood easily once you know how to interpret it.

In this investigation, you will:

- create graphs of velocity versus position and time.
- create a predictive model for the velocity of a cart rolling down a hill.

Materials List
- DataCollector
- SmartTrack
- Energy Car
- Photogate
- Physics Stand
- Velocity sensor

1 Velocity measurements

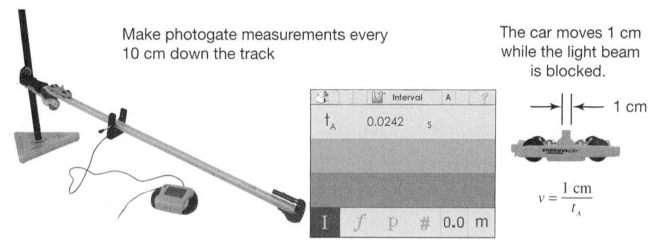

Make photogate measurements every 10 cm down the track

The car moves 1 cm while the light beam is blocked.

t_A 0.0242 s

$v = \dfrac{1 \text{ cm}}{t_A}$

1. Set up the track with the velocity sensor attached. Use the fifth position from the bottom of the stand. Set the DataCollector to Timer mode and plug one photogate into input A.
2. Place the photogate 10 centimeters from the top of the track. This will be the location of your first trial. You will perform several trials by moving the photogate further down the track in 10-centimeter increments for each new trial. Record the photogate position and the time it takes the 1 centimeter flag to break the light beam for each trial you perform in Table 1.

Table 1: Time and velocity data

Photogate position (cm)	Time through A (s)	Velocity (m/s)

2 Analyzing data

a. Calculate the velocity of the car in cm/s, and record the results in Table 1.

b. Do the velocities increase or stay about the same along the ramp?

c. Repeat the time measurement five times at one of the locations you measured. Calculate the velocity for each of the five trials. How precise is your data compared to that of other groups? Justify your conclusion by comparing the estimated error of each group.

Investigation 2.2 Position, Velocity, and Time Graphs

3 The velocity vs. time graph

In the next part of the investigation, you are going to use the graphing capability to generate the graph of velocity versus time for the track, and compare your data with the data you collected with the photogate.

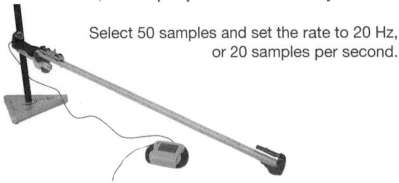

Select 50 samples and set the rate to 20 Hz, or 20 samples per second.

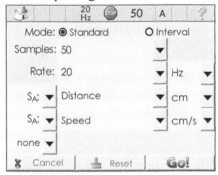

1. Switch the cord from the photogate to the velocity sensor and remove the photogate. Without changing the angle of your ramp, start a new experiment in Data Collection mode. Select the Setup tab at the bottom of the screen and set the samples to 50 and the rate to 20Hz. Click back to Meter view to see the time, distance, and velocity.
2. Attach the target to the back of the car.
3. Put the car at the top of the track, hit the Go button, and release the car.

4 Analyzing and interpreting data

a. Once the DataCollector takes all 50 samples, switch to Table view using the Table tab on the menu bar at the bottom of the screen.

b. Scroll through the position, velocity, and acceleration data. Set the time column to be X and the velocity column to be Y1. Switch to Graph view.

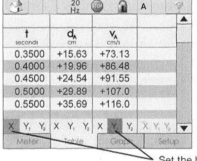

Set the horizontal axis to be t, and the vertical axis to be v

c. Describe the graph you see. Is it a straight line or a curve? Does it get steeper with time or shallower?

d. Is there acceleration in the motion of the car? How do you know from the velocity vs. time graph?

e. Go back to table view and set position to be X and velocity to be Y1. Return to Graph view and observe the graph you see. Is it a straight line or a curve? Does it get steeper with time, or shallower?

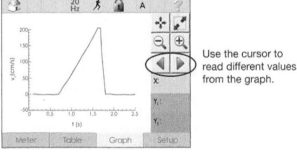

Use the cursor to read different values from the graph.

f. Move the cursor to a position on the graph that is as close as possible to a position where you made a photogate measurement of velocity. Write down the value of both position and velocity. (NOTE: To compare values, you must subtract 6.7 centimeters from the position to find the velocity at the same location for the car.)

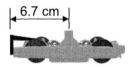

6.7 cm

g. Compare the graph values with the photogate calculation you did at the same position in Part 2.

h. Derive a formula that will give you the photogate time if you know the velocity and the distance traveled. Use the cursor to choose a point on the graph, and test your formula. (Note: don't forget that you need to consider when in time your car started moving.)

14

2.3 Accelerated Motion

What happens to the velocity of an object as it moves downhill?

An object that is changing velocity is accelerating. By measuring an object's velocity and examining how its velocity changes, the object's acceleration can be calculated. In this investigation, you will use two methods to calculate the acceleration of a car rolling down a track.

Materials List
- DataCollector
- 2 Photogates
- Velocity sensor
- SmartTrack
- Energy Car
- Calculator

1 Setting up the experiment

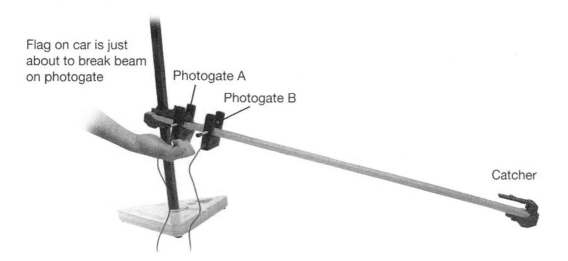

1. Attach the SmartTrack to the sixth hole from the bottom of the physics stand.
2. Position photogate A so it is attached at the 10-centimeter mark of the SmartTrack, and position photogate B at the 20-centimeter mark.
3. Connect both photogates to the DataCollector.
4. Select Timer mode on the DataCollector's Home screen.
5. Select Interval function by tapping the "I" icon. Interval function is the default function in Timer mode, but check the menu bar at the bottom of the screen to ensure that the "I" icon is illuminated. The status bar at the top of the screen will say Interval.

2 Collecting data

1. Place the car with no marbles in it at the top of the SmartTrack and use the hold button on the underside of the SmartTrack to keep the car in place.
2. Release the hold button to allow the car to roll down the ramp.
3. Record the time (t_A) it takes the car to move through photogate A in Table 1.
4. Record the time (t_B) it takes the car to move through photogate B in Table 1.
5. Record the time (t_{AB}) it takes the car to move from A to B in Table 1.

Investigation 2.3 Accelerated Motion

6. Move photogate B 10 centimeters down the ramp to the 30-centimeter mark. Release the car in the same manner as you did for the previous step. Record t_A, t_B, and t_{AB} for each trial in Table 1.
7. Repeat for a total of eight different A-to-B distances as shown in Table 1.

Table 1: Distance and time data

Distance from A to B (cm)	Time through photogate A t_A (s)	Time through photogate B t_A (s)	Time from A to B t_{AB}(s)
10			
20			
30			
40			
50			
60			
70			
80			

3 Analyzing your data

Table 2: Velocity and acceleration

Distance from A to B (cm)	Velocity through photogate A (cm/s)	Velocity through B (cm/s)	Acceleration (cm/s^2)
10			
20			
30			
40			
50			
60			
70			
80			

a. Use the time through photogate A and the length of the flag on the top of the car (1 centimeter) to calculate the velocity of the car when it passed through photogate A for each trial. Record your result in Table 2.

b. Use the time through photogate B and the length of the flag on the top of the car (1 centimeter) to calculate the velocity of the car when it passed through photogate B for each trial. Record your result in Table 2.

c. Use the velocity at A, the velocity at B, and the time it took the car to move down the ramp (t_{AB}) to calculate the car's acceleration for each trial. Write out a sample calculation to show how you found the acceleration. Record your result in Table 2.

d. Were the calculated accelerations roughly the same, or were they different? What does this tell you about the car's velocity as it rolled down the ramp?

4 Making and analyzing a graph

a. Make a velocity vs. time graph using your data. Plot the time from A to B on the *x*-axis and the velocity at B on the *y*-axis. Draw a best-fit line and be sure to label the axes and title the graph.

b. The slope of a line is found by dividing the rise (vertical change) by the run (horizontal change). What is the meaning of the slope of your velocity vs. time graph?

c. Should your graph show a straight line or a curve? What would be the difference between the two?

d. How does your answer to the previous question relate to the acceleration of the car that you calculated at the eight different locations on the SmartTrack?

e. Calculate the slope of your best fit line. Use the end points of your best fit line to calculate its overall slope. If the line made by your data is essentially straight, just use the velocities you calculated and the t_{AB} data you collected to determine the line's slope.

f. How does this value compare to the average of all the accelerations you calculated in Table 2?

g. Suppose the car had rolled along the ramp at a constant velocity. What would your velocity vs. time graph have looked like?

h. Suppose the car had decreased in velocity while rolling along the ramp. What would your velocity vs. time graph have looked like?

5 *Challenge*: Analyzing your data for error

a. What additional equipment and sensors do you have that may allow you to check the work you have done in this investigation up to this point? Describe the process you would follow to check the data you collected in Part 2, and the calculations you made in Part 3.

b. How could you check the graph you made in Part 3? Describe the process you would follow to check your graph.

c. Follow your own steps and perform the checking activities you came up with above for your data, your calculations, and your graph.

d. Describe how your data, calculations, and graph compare to the results you saw in your checking activities.

e. Discuss the possible causes of any discrepancies you may have found during your checking activity.

f. Is there anything you could have done differently to improve the accuracy or precision of your data collection in Part 2?

Investigation 3.1 The Law of Inertia

Why are heavier objects harder to start moving or stop from moving?

This investigation is about mass and inertia. Inertia is the property of an object that resists changes in motion. Inertia comes from mass. Objects with more mass have proportionally more inertia. In this investigation, you will explore Newton's first law, the law of inertia.

Materials List
- Energy car
- SmartTrack
- 3 Steel marbles
- Graph paper
- DataCollector
- Photogate
- Electronic balance or triple-beam balance
- Physics stand

1 Launching cars of different mass

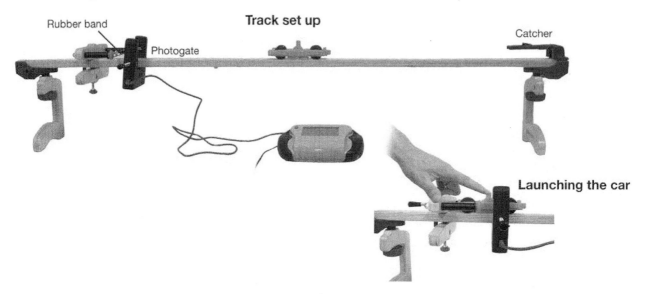

1. Set up the SmartTrack with a launcher at the 20-centimeter mark facing the catcher end.
2. Put a rubber band onto the arms of the launcher in an *x* pattern by giving the rubber band a half-twist, then secure it in place between the retaining washers with the securing screws.
3. Put one photogate about 10 centimeters away from the rubber band. Plug it into the photogate A slot on the DataCollector. Use the Timer mode and the Interval function to measure the time the car's flag breaks the photogate's beam.
4. Use the plunger screw to set one constant deflection of the rubber band. This means each time a car is launched, the same force is applied to each car.
5. Perform four trials by launching cars with 0, 1, 2, and 3 steel marbles. Record the mass and the time through the photogate for each launch in Table 1. Calculate and record the speed in Table 1.

Table 1: Constant force data

Mass of car (kg)	Time through photogate (s)	Speed (m/s)

2 Constructing explanations

a. Use Table 1 to graph the speed of the car (y) against the mass (x).

b. Why did the speed change when the same launching force from the rubber band was applied to cars of different mass? How do your observations support your answer?

3 Inertia and weight

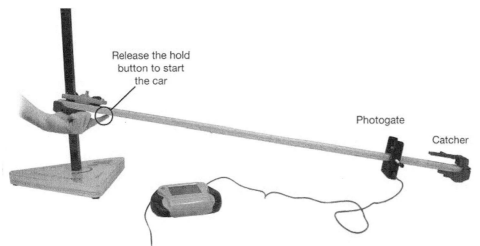
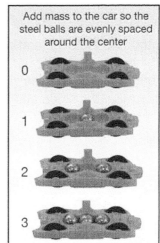

1. Attach the SmartTrack to the fourth hole up from the bottom of the stand with a blue-threaded knob. Attach the photogate to the 100-centimeter mark.
2. Use the hold button at the top of the track to drop the car from the same place each time.
3. Drop cars of four different masses—with 0, 1, 2, and 3 steel marbles—from the same height on the hill. Use the DataCollector to measure the time through the photogate, and record it in Table 2.
4. Record the mass and calculate the speed of each trial in Table 2.

Table 2: Constant height data

Mass of car (kg)	Time through photogate (s)	Speed (m/s)

4 Arguing from evidence

a. What force makes the car go down the hill? What property of matter does this force act upon?

b. Does increasing the mass of the car increase its speed by a proportional amount? Does the speed decrease with increasing mass? Or does the speed stay about the same, no matter what the mass?

c. Propose and discuss an explanation for why changing the mass has a very different effect on the speed when gravity provides the force compared to when the force is provided by a rubber band.

d. Research and define the terms *inertia*, *weight*, and *mass*. Write two to three sentences that describe how these three concepts are similar and how they are different.

3.2 The Second Law: Force, Mass, and Acceleration

What is the relationship between force, mass, and acceleration?

British scientist George Atwood (1746–1807) used two masses on a light string running over a pulley to investigate the effect of gravity. You will build a similar device, aptly called an Atwood's apparatus, to explore the relationship between force, mass, and acceleration.

In this investigation, you will:

- determine the acceleration for an Atwood's apparatus of fixed total mass.
- create a graph of force versus acceleration for the Atwood's apparatus.
- determine the slope and y-intercept of your graph and relate them to Newton's second law.

Materials List

- Physics Stand
- Atwood's Apparatus
- Red safety string
- 2 Mass hangers
- Steel washers (18)
- Plastic washers (6)
- Measuring tape
- Photogate
- DataCollector
- Electronic scale or triple beam balance

1 Analyzing the Atwood's apparatus

To accelerate a mass, you need a net force. Newton's second law shows the relationship between force, mass, and acceleration:

NEWTON'S SECOND LAW

$$F = ma$$

Force (N), Mass (kg), Acceleration (m/s²)

The Atwood's apparatus is driven by an external force equal in magnitude to the weight difference between the two mass hangers.

$$F_{weight} = (m_1 - m_2)g$$

You will vary the two masses, m_1 and m_2, but you will keep the total mass constant. As you move plastic washers from m_2 to m_1, you will use a photogate to measure the acceleration of the system. If you know the acceleration and the total mass of the system, you will be able to calculate the applied force that is responsible for accelerating the system. The equation for the system's motion is a variation of the basic second law formula:

$$F_{applied} = (m_1 + m_2)a$$

An ideal pulley would be frictionless and massless, and would just redirect the one-dimensional motion of the string and attached masses without interfering with the motion. However, the pulley you will use has mass and there will be some friction involved. For the purpose of this investigation, we will neglect the mass of the pulley, but we will be able to analyze the friction ($F_{friction}$) involved with our Atwood's apparatus. To represent the friction present in the system, you must subtract it from the force of weight, since the friction opposes this force.

$$F_{applied} = F_{weight} - F_{friction}$$

The Second Law: Force, Mass, and Acceleration   **Investigation 3.2**

2 Setting up the Atwood's apparatus

1. Set up the Atwood's apparatus as shown in the photograph. Attach the pulley to the top of the physics stand.
2. Attach the mass hangers to the red safety string. Place eight steel washers and six plastic washers on one mass hanger. This will be m_2. Place 10 steel washers on the other mass hanger. This will be m_1. Place the string over the pulley.
3. Pull m_2 down to the stand base. Place a sponge or some other small cushion on the base to protect it from the falling m_1. Let go of m_2 and observe the motion of the Atwood's apparatus.

a. Which mass moves downward and why?

b. What would happen if m_1 and m_2 were equal masses? Why?

c. Do the masses accelerate when they move? Explain.

d. How does the acceleration of m_1 compare to the acceleration of m_2?

e. The force responsible for moving the weights equals the weight difference of the mass hangers. Write the simple formula that will allow you to use the mass difference and g, the acceleration due to gravity, to calculate the weight difference of the mass hangers (F_{weight}).

3 Collecting data

1. Find the total mass of m_1 and m_2 and record in Table 1. This is the same for each trial.
2. Attach a photogate to the Atwood's apparatus as shown. Plug the photogate into the DataCollector (input A). As the pulley rotates, the striped pattern on the pulley will break the light beam of the photogate.
3. Turn on the DataCollector. At the Home window, select Data Collection.
4. At the Go window, tap on the Setup option at the bottom of the screen.
5. In the Setup window, choose Standard mode.

6. Set samples to 500 and the data rate to 5 Hz. For input, select the SP_A setting, then select speed in cm/s. Set the other inputs to none, working from bottom to top.
7. Measure the mass difference between m_2 and m_1. Record in Table 1.
8. Pull m_2 to the base. Tap Go at the bottom of the Setup window.
9. When the experiment has started, release m_2. When the hanger falls onto the cushion, press the button on the DataCollector enclosure to stop the experiment.

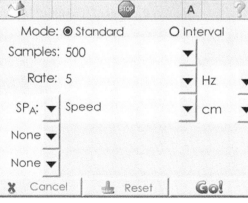

10. Select the Graph tab at the bottom of the screen, and study the graph of the speed data. The speed should be increasing at a constant rate. Determine the acceleration by calculating the slope of the

21

Investigation 3.2 The Second Law: Force, Mass, and Acceleration

speed versus time graph. Convert to units of m/s². Record the acceleration in Table 1. Calculate $F_{applied}$ and record it in Table 1. There is a formula reminder provided in the table.

11. Calculate F_{weight} (see your answer to 2e) and enter your result in Table 1.
12. Transfer *one* of the plastic washers from m_2 to m_1.
13. Press the button on the DataCollector enclosure to resume data collection. Repeat steps 7–12 until you have transferred all of the plastic washers to m_1, one at a time.

Table 1: Acceleration, mass, and force data

Total mass $(m_1 + m_2)$ (kg)	Mass difference $(m_1 - m_2)$ (kg)	F_{weight} $(m_1 - m_2)g$ (N)	Acceleration (m/s²)	$F_{applied}$ $(m_1 + m_2)a$ (N)	$F_{friction}$ $(F_{app} - F_w)$ (N)

4 Arguing from evidence

a. Graph F_{weight} versus acceleration (F_{weight} on the y-axis and acceleration on the x-axis). Draw a best-fit line.

b. What kind of relationship does the graph show? Is this consistent with Newton's second law? Explain.

c. Determine the slope of your line. What is the significance of the slope in your experiment?

d. Compare your slope with the total mass of your system. What is the percent difference? What could account for any difference?

e. Determine the y-intercept of your line. The equation for a line is $y = mx + b$ (m is the slope and b is the y-intercept). Substitute the variables in for y, m, and x. Compare this equation to the one presented in part 1.

f. Based on your answer to the previous question (4e), what does the y-intercept represent? Does this value make sense? Explain.

g. Use the formula below to calculate $F_{friction}$ for each trial and record your results in Table 1. How do these calculated values compare to your graph's y-intercept?

$$F_{friction} = F_{applied} - F_{weight}$$

3.3 Free Fall

What kind of motion is falling?

What kind of motion is falling? We know falling objects accelerate, and that is a good place to start when examining their motion.

In this investigation, you will:

- determine an equation for the velocity in free fall.
- use the equation to make predictions.

Materials List
- DataCollector
- Physics Stand
- SmartTrack
- Steel marbles (3)
- Velocity sensor
- Energy car with magnet plate
- Bubble level
- Rag, soft cloth, or sponge

1 Free fall

1. Set up the physics stand and the velocity sensor from the SmartTrack as shown in the diagram, putting the stand on the floor. Putting a rag in the catcher helps to eliminate noise in the data.

2. With the DataCollector on a table, select Data Collection mode from the Home screen and use the setup tab to set the samples to 20 and the rate to 20 Hz.

3. Attach the target to the EnergyCar, press the hold button, and use the target to hang the car from the pin that comes through the track. Make sure the wheels are properly set on the track.

4. Press the Go button and release the hold button after the countdown. The DataCollector will measure the time, position, and velocity of the car as it falls.

5. The fall takes less than one second. Be careful not to push or shake the track. To get the technique right, you should perform several trials.

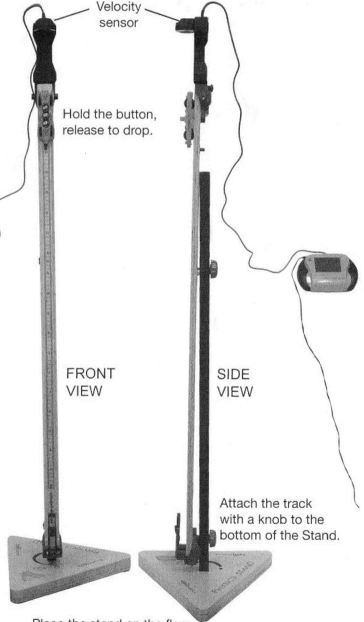

23

Investigation 3.3 Free Fall

2 Analyzing the data

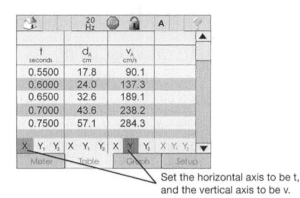

Set the horizontal axis to be t, and the vertical axis to be v.

a. Once the DataCollector takes the samples, switch to Table view. You should be able to scroll through the position and velocity data.

b. Set the time (*t*) column to be X and set the velocity (*v*) column to be Y1. Switch to Graph view.

c. Describe the graph you see. Is it a straight line or a curve?

d. Which quantity of this experiment is represented in the slope of your graph? Calculate the slope from the velocity and time at two separate points on the graph. (*Note*: The positive direction is defined as the direction moving away from the sensor, so positive motion indicates that the marble's direction of movement is *down* in this experiment.)

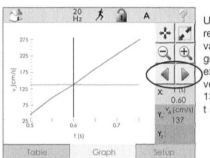

Use the cursor to read different values from the graph. In the example, the velocity is 137 cm/s at $t = 0.60$ s.

e. The equation for a straight line is $y = mx + b$. Determine the slope, *m*, and the y-intercept, *b*, from the graph using the cursor. Rewrite the equation for a straight line using the following variables: velocity, *v*; initial velocity, v_0; acceleration due to gravity, *g*; and time, *t*.

AVERAGE ACCELERATION

$$g_{avg} = \frac{v_2 - v_1}{t_2 - t_1}$$

f. Calculate the acceleration (slope) from two points in the 30-to-60-centimeter range of the track. Compare this with the acceleration you derive from the slope of the line in d above. Are the two values similar? By what percentage are they different? Give a possible explanation for any differences.

3 Different masses

1. Repeat the experiment with the car and one or two steel marbles. You need the magnet plate to hold the marbles in during the fall. Careful, the marbles can fall out on impact. The car and target are about 58 grams and the steel marbles are 28 grams. You can also try other objects. Not all objects will work. Some objects might not be reflective enough to work with the sensor.

4 Using a model to construct explanations

a. How do your values of acceleration compare with the accepted value of *g*, which is 9.82 m/s^2? Your answer should be in terms of a percent difference between your result and the accepted value.

b. Suppose friction from the air was substantial enough to observe. What would the effect be? Would air friction tend to increase or decrease the acceleration you measured?

c. From your multiple trials, you should be able to estimate an error for your measurement. How big is this error relative to the difference you calculated in a above?

d. Based on b above, do you find that your experiment is in agreement with the accepted value of *g*?

e. Discuss any possible sources of error in the experiment. Make sure you assess whether each potential source of error could be large enough to explain any differences you observed.

4.1 Newton's Third Law and Momentum

What happens when equal and opposite forces are exerted on a pair of Energy Cars?

When you apply a force to throw a ball, you also feel the force of the ball against your hand. That is because all forces come in pairs called *action* and *reaction*. This is Newton's third law of motion. There can never be a single force (action) without its opposite (reaction) partner. Action and reaction forces always act in opposite directions on two different objects. In this investigation, you will set up two Energy Cars to study Newton's third law.

Materials List
- 2 Energy Cars
- SmartTrack
- DataCollector
- 2 Photogates
- Rubber band
- 4 Steel marbles
- Energy Car link
- Bubble level

1 Setting up and starting the experiment

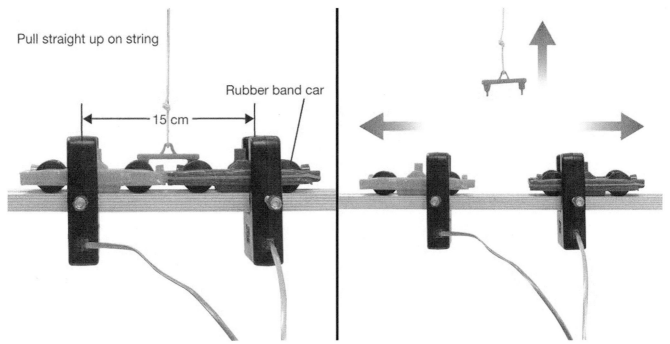

1. Set up the SmartTrack with its feet attached so it sits horizontally on your lab table.
2. Use the bubble level to get the SmartTrack level. You may need to slide a few sheets of paper under one of the feet to make the entire SmartTrack level.
3. Check with the bubble level in several locations, including the middle and all along the length of the SmartTrack, to make sure it is level.
4. Attach the launcher on the end opposite the catcher. Place a small ball of clay on the launcher's plunger screw. This will stop the car moving in that direction from bouncing back down the track.
5. Place one steel marble in each car, and wrap one car with a rubber band.
6. Place two photogates 15 centimeters apart as shown in the photo.
7. Place the two cars "nose-to-notch" between the photogates.
8. Squeeze the cars together and attach them with the Energy Car link.

Investigation 4.1 Newton's Third Law and Momentum

9. Center the attached car pair between the photogates so each is about to break the photogate's beam, but do not actually break the beam. Check that both photogates' indicator lights are still green. Make sure all four wheels of both cars are on the track.
10. With a *very quick* upward motion, pull the link *straight up* and out from the cars. **CAUTION: Wear eyeglasses or safety glasses to avoid injury.**
11. Observe the time through each photogate. Repeat several times.

2 Using your model

a. What caused the two cars to move when you took out the link?

b. According to Newton's third law, the cars experienced equal and opposite forces. How can you tell this is true by looking at the time through each photogate?

c. If one car was twice the mass of the other, would the cars still experience equal and opposite forces? Why or why not?

3 Changing the masses

Four combinations

NOTE: Adding two steel marbles to the Energy Car doubles its mass.

1. Try the experiment with the four combinations of mass shown to the right. Take the average of three trials for each and record your data in Table 1. *CAUTION: Wear eyeglasses or safety glasses to avoid injury.*
2. Calculate the average speed for each trial and record in Table 1.

Table 1: Energy car action/reaction data

Marble pairings for connected cars		Time through photogate (s)		Speed (cm/s)	
A	B	A	B	A	B
0 marbles	2 marbles				
0 marbles	0 marbles				
2 marbles	0 marbles				
2 marbles	2 marbles				

4 Arguing from evidence

a. How does the speed of each car pair compare when masses are equal?

b. How does the speed of each car compare when one of the pair has twice the mass?

c. Explain how your speed data supports the idea that there are equal and opposite action and reaction forces acting on the cars.

d. If the action and reaction forces are equal in strength, when the cars separate, why does one car move at a speed different from that of the other car when the masses are unequal? (*Hint*: The answer involves the second law of motion.)

4.2 Conservation of Energy

How can you predict the maximum velocity of a pendulum?

As a pendulum swings, where does the pendulum bob have the greatest potential energy? The least potential energy? Where is the kinetic energy the greatest? Where does the bob have the least kinetic energy? The swinging pendulum's potential-to-kinetic energy trade-offs are a direct result of the law of conservation of energy. In this investigation, you will use the law of conservation of energy to predict the pendulum bob's maximum velocity at four different release heights.

Materials List
- Physics Stand
- 2 Photogates
- 2 Threaded knobs
- DataCollector
- Red string with lanyard clip and cord lock
- Pendulum bob assembly

1 The pendulum's energy

1. Assemble the pendulum bob as shown in the photo. Fasten the hex nut to the very bottom of the threaded rod and slide the pendulum tube over the rod.

2. Fasten a photogate to the top and bottom of the physics stand. The upside-down photogate at the top of the stand serves as an attachment point for the pendulum string. You will not be using this photogate for data collection. Slide the red string up through the hole in the middle of the gate and fasten a cord lock to the end of the string to keep the string from falling back through the hole.

3. Pull the pendulum back a short distance and see if you can release it so it swings freely though the bottom photogate. Practice several times until you can get consistent releases. Watch the pendulum as it swings back and forth.

a. Where in its swing does the pendulum have the highest *potential energy* (E_p)? Where does it have the lowest E_p?

b. Where in its swing does the pendulum have the highest *kinetic energy* (E_k)? Where does it have the lowest E_k?

c. Based on your answer to question b, where does the pendulum bob have maximum velocity?

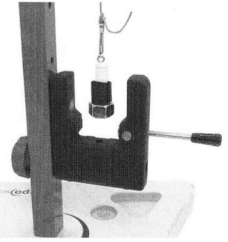

According to the law of conservation of energy, the potential energy of the pendulum before it is released equals the kinetic energy of the pendulum at the bottom of the swing. Therefore, $E_{p\ initial} = E_{k\ final}$. This equation and simple substitutions show how you can predict the maximum velocity of the pendulum bob.

Investigation 4.2 Conservation of Energy

2 Predicting the maximum velocity

What is the pendulum bob's maximum velocity when it is released from a height that is 0.15 meters higher than its original height? To answer this question, use the formula that solves for the pendulum's maximum velocity. The height you will use in the calculation is the change in height, or Δh. These have been filled in for you.

$$E_{p\ initial} = E_{k\ final}$$

$$E_p = mgh; \quad E_k = \frac{1}{2}mv^2$$

$$mgh = \frac{1}{2}mv^2; \quad v = \sqrt{2gh}$$

1. Use the formula for the maximum velocity of the pendulum bob to complete Table 1. These are your velocity predictions. Is energy really conserved as the pendulum swings back and forth? If so, the experiment you do in Part 5 should give you velocity values very close to your predictions.

Table 1: Predictions for maximum velocity

g (m/s^2)	Δh (m)	Maximum velocity (m/s) $v = \sqrt{2gh}$
9.8	0.15	
9.8	0.20	
9.8	0.25	
9.8	0.30	

3 Setting up the pendulum

1. The pendulum tube is what breaks the photogate's beam. The diameter of the pendulum tube, 0.020 m, is the distance the bob travels as it passes through the photogate. *Use this distance in all velocity calculations.*

2. Align the pendulum bob so the tube is the only part of the bob that breaks the photogate's beam.

3. The point where the pendulum tube and the hex nut meet is the center of mass of the pendulum bob. Measure the distance from the center of mass of the pendulum bob to the lab table (or floor if the physics stand rests on the floor). This is your starting height, or h_1.

4. Turn on the DataCollector. Select CPO timer, and then Interval mode.

5. Connect the bottom photogate to the photogate A port on the back of the DataCollector.

6. Swing the pendulum through the photogate a few times to make sure the pendulum bob breaks the photogate beam correctly.

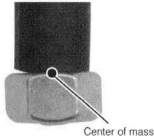

Center of mass

4 Creating a height reference scale

Raising the pendulum bob to each new height is much easier if you have a reference scale to help you set the release heights.

1. Divide a sheet of copy paper or card stock in half the long way, and tape the two pieces together, end-to-end.
2. Starting from one end of the paper strip, measure a distance equal to h_1 (from Part 3, step 3). Draw a height reference line straight across the paper strip. Label this line h_1.
3. You will raise the bob 0.15 meters from h_1 for your first five trials, 0.20 meters from h_1 for your second five trials, and so on. Measure a distance of 0.15 meters from h_1, draw a height reference line straight across the paper strip, and label it $\Delta h = 0.15$ m.
4. In a similar way, draw height reference lines for the remaining three Δh values.
5. Tape the paper scale to the front of a meter stick. The bottom of the paper strip should be flush with the end of the meter stick touching the floor.
6. One person will hold the height reference scale vertically behind the bob. The pendulum releaser will hold the bob and align the center of mass of the bob just above the hex nut with the appropriate Δh reference line. This is how you will get accurate release heights for your pendulum.

5 Testing your predictions

1. Raise the bob to the first Δh value marked on the reference scale.
2. Let the pendulum bob swing through the photogate. Record the time from photogate A in Table 2.
3. Repeat steps 1 and 2 four times so you have a total of five trials at this height.
4. Calculate the velocity for each trial and record in Table 2.
5. Calculate the average velocity and record in Table 2.
6. Repeat steps 1–5 for the remaining Δh values.

Table 2: The pendulum's velocity

Δh (m)	0.15		0.20		0.25		0.30	
	Time (s)	Velocity (m/s)	Time (s)	Velocity (m/s)	Time (s)	Velocity (m/s)	Time (s)	Velocity (m/s)
Trial 1								
Trial 2								
Trial 3								
Trial 4								
Trial 5								
Average								

Investigation 4.2 Conservation of Energy

6 Analyzing the data

1. How close were your predictions? Copy your velocity predictions from Table 1 into Table 3. Copy the measured velocities from Table 2 into Table 3.
2. Calculate the percent error at each Δh value and record in Table 3.

Table 3: Comparing predicted velocities to measured velocities

Δh (m)	Predicted velocity (m/s)	Measured velocity (m/s)	% error $\left\|\dfrac{(\text{measured}-\text{predicted})}{\text{measured}}\right\| \times 100$
0.15			
0.20			
0.25			
0.30			

a. How close were your predictions to the actual velocities? Discuss.

b. What part of your procedure left the most room for procedural error? Explain.

c. You began this investigation by assuming that the pendulum's motion obeys the law of conservation of energy. Did you find this to be true? Discuss.

4.3 Collisions

Why do things bounce back when they collide?

Newton's third law tells us that when two objects collide, they exert equal and opposite forces on each other. However, the effect of the force is not always the same. What happens when you collide two Energy Cars that have unequal masses? In this investigation, you will perform several collisions with cars of different masses and compare the results.

Materials List
- 2 Energy Cars
- SmartTrack
- DataCollector
- 2 Photogates
- 2 Rubber bands
- Electronic scale or triple beam balance
- 4 Steel marbles
- Bubble level

1 Modeling a collision

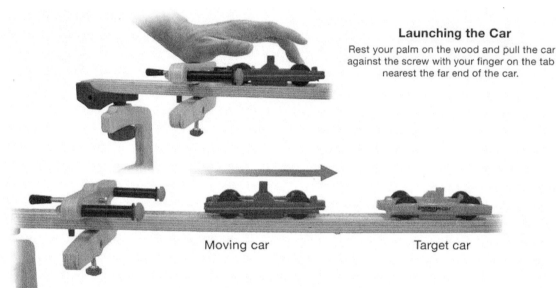

Launching the Car
Rest your palm on the wood and pull the car against the screw with your finger on the tab nearest the far end of the car.

Moving car   Target car

1. Set up the SmartTrack with the launcher on the end opposite the stopper.
2. Put a rubber band onto the arms of the launcher in an *x* pattern by giving the rubber band a half-twist, then secure it in place between the retaining washers with the securing screws.
3. Use the bubble level to set the track level.
4. Place one steel marble in each car.
5. Wrap a rubber band around the moving car. Place both cars on the track so their noses are pointed toward the rubber band launcher.
6. Place the target car near the center of the track. Use the plunger screw to launch the car using the same deflection of the rubber band each time. This means the same force is applied to each launch. You will use this car to create the collision.
7. To make a collision, release the moving car from the launcher once you have pulled it back and stretched the rubber band so the car rests against the flat face of the plunger screw.
8. The moving car will speed down the track and hit the target car. This is an efficient way to produce collisions on the track.

Investigation 4.3 Collisions

a. Does the *moving* car bounce back after the collision?
b. Does the *moving* car keep going forward after the collision?
c. Does the *moving* car stop at the collision?
d. How does the *target* car behave?

2 Using your model

a. Describe the motion of the two cars before and after the collision.
b. The target car must exert a force on the moving car to stop it. How strong is this force relative to the force the moving car exerts on the target car to get it moving? How could you use the photogates to provide evidence for your answer?

3 Collecting evidence

Place photogates 20 cm apart at the middle of the SmartTrack

1. Try the experiment again, but now use two photogates to collect time data.
2. Place two photogates 20 centimeters apart at the middle of the track. Plug the one closest to the moving car into the photogate A slot, and the other into the photogate B slot of the DataCollector.
3. Put the target car on the track so it is very close to, but not quite breaking the beam of, photogate B.
4. Release the moving car with the rubber band from the launcher as before and make a collision.
5. Repeat several times and record trial times in Table 1.

Table 1: Collision times

Collision Trial	Time for moving car to pass through A before collision (s)	Time for target car to pass through B after collision (s)
1		
2		
3		
4		
5		

a. Newton's third law tells us that when the moving car exerts a force on the target car, the target car exerts an equal and opposite force on the moving car. Does your data provide evidence for this? Explain.
b. You can compare times through A and B for each individual trial. How can using these times show there are equal and opposite forces at work?

Collisions   Investigation   4.3

4 Changing the masses

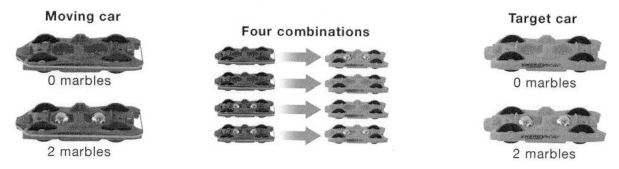

1. Try the experiment with the four combinations of mass shown above. You do not need to use photogates for this part of the investigation.

5 Arguing from evidence

a. Describe the motion of the two cars when the target car has more mass than the moving car.

b. Describe the motion of the two cars when the target car has less mass than the moving car.

c. Explain how your observations support the idea that there are action and reaction forces.

d. If the action and reaction forces are equal in strength, why do the two cars move at different speeds after the collision when the masses are unequal? (*Hint*: The answer involves the Newton's second law.)

5.1 Working with Force Vectors

How can we use force vectors to predict the acceleration of the car?

If an object is in equilibrium, all of the forces acting on it are balanced, the net force is zero, and the object does not experience any acceleration. If the forces are not balanced, there *is* a net force and the object *does* experience acceleration. With careful observation, measuring, and calculations, unknown forces can often be figured out by using vectors. In this investigation you will calculate the net force on the Energy Car and use force vectors to predict its acceleration.

Materials List
- Physics Stand
- DataCollector
- SmartTrack
- Velocity sensor
- 3 Steel marbles
- Graph paper
- Energy Car
- Protractor
- Metric ruler
- Electronic scale or triplebeam balance\
- Pencil

1 What is a force vector?

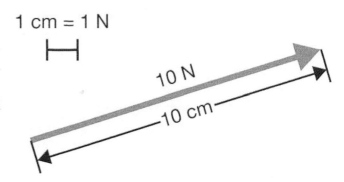

A force vector includes both the force's magnitude and direction. The magnitude of a force vector is expressed in newtons. When drawn to scale, the length of a force vector corresponds to a certain amount of force. Scaled vectors like the one shown can give you useful information. In this investigation, you will draw your own scaled force vectors.

2 The components of a force vector

When you draw a force vector on a graph, distance along the x- or y-axes represents the strength of the force in the x- and y-directions. A force at an angle has the same effect as two smaller forces aligned with the x- and y-directions. In the figure to the right, the 8.7-N and 5-N forces applied together have the same effect as a single 10-N force applied at an angle of 30° from the y-axis. This idea of breaking a force down into its x- and y-components is very important, as you will see.

a. If you place the car on the SmartTrack when it is on a flat, level table, what forces are acting on the car? Would the car move?

b. In what directions would the forces be acting?

c. What happens to those forces when you put the car on the SmartTrack when it is at an angle? Does the car move? Why?

1. Use a protractor to find the correct angle, then draw the force vector to scale.

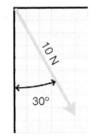

2. Extend lines to the x- and y- axes.

3. Read off the x- and y- components

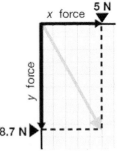

3. The angle of the ramp and the force of weight (F_w)

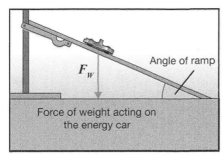

Force of weight acting on the energy car

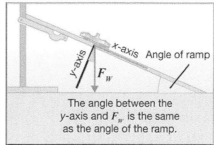

The angle between the y-axis and F_w is the same as the angle of the ramp.

Your ramp will be at an angle of 15° when you attach it to the fifth hole from the bottom of the stand. The force of weight (F_w) always points straight down for any object, from the center of mass of the object straight toward the center of Earth. This means that there is a component of F_w that is parallel to the ramp, which we will call F_{ramp}. There is also a component that is perpendicular to the ramp, called the normal force (F_{normal}), which doesn't help move the car down the ramp, so we will concentrate on the one that does.

a. What would a force vector acting parallel to the ramp do to the car?

b. What about the vector would tell you in which direction it was acting on the car?

c. What about the vector would tell you how much force was being applied to the car?

4. Finding F_w

You will need Newton's second law to find the F_w of your Energy Car.

NEWTON'S SECOND LAW

$$F = ma$$

Force (N), Mass (kg), Acceleration (m/s²)

You want to know a force, so you need to determine the mass and the acceleration and then solve for the force.

1. Measure and record the mass of the Energy Car in Table 1. Be sure to record the mass in kg.
2. The acceleration you need is the acceleration of gravity: 9.8 m/s².
3. Use Newton's second law to determine the F_w of the energy car, and record your result in Table 1.

Table 1: Mass, acceleration, and force data

Mass of car (kg)	Acceleration of gravity (m/s²)	F_w Force of weight (N)

Investigation 5.1 Working with Force Vectors

5 Resolving F_w into its component force vectors

F_w is a force vector, and can be broken down into component vectors as they apply to the ramp. The key pieces of information you need to know are

- the angle of the ramp.
- the direction of the force vector F_w.
- the magnitude of the force vector F_w.

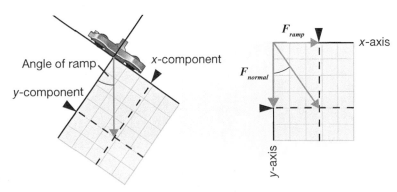

a. Fill in Table 2 with the information you need. Leave the fourth column blank for now.

Table 2: F_w component resolution data

Angle of the ramp (degrees)	Direction of F_w	Magnitude of F_w (N)	Magnitude of F_{ramp} (N)

b. Use a pencil, a protractor, a ruler, and a piece of graph paper to resolve F_w into its component vectors and find F_{ramp}. Be sure to decide on a scale for your vector drawing, label the axes, and label all the force vectors. When you find the magnitude of F_{ramp}, record it in Table 2.

6 Using F_ramp to make a prediction

Now that you have found F_{ramp}, you will use that piece of information and whatever else you need to predict the acceleration of the car on the ramp.

a. Other than F_{ramp}, what other piece of information will you need to predict the acceleration of the car?
b. What formula did you use to find F_w?
c. How can you use that formula with F_{ramp} to predict the acceleration of the car down the ramp?
d. Perform your calculation and state your predicted acceleration. Enter the value into Table 3.

7 Test your prediction

1. Attach the velocity sensor to the SmartTrack
2. Attach the spur to the SmartTrack, then attach the spur to the fifth hole up from the bottom of the physics stand.
3. Connect the sensor to the DataCollector with the cable.
4. From the home screen of the DataCollector, select Data Collection, then tap Setup.

5. Set the samples to 50 and the rate to 20 Hz. For inputs, select speed as the only input. You may need to de-select any additional inputs, by changing extra inputs to "None."

6. Put the car at the top of the ramp and keep it in place using the hold button.

7. When your group is ready, tap the Go screen, and allow the car to roll down the ramp by releasing the hold button.

8. Go to the graph screen by tapping the Graph tab at the bottom of the display.

9. Acceleration is the slope of the velocity vs. time graph. Select two points from the graph where the car was in motion to determine the acceleration of the car.

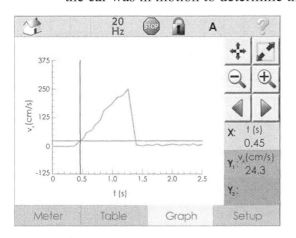
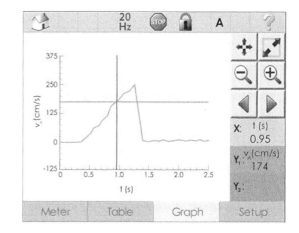

10. Record the slope in Table 3 as the measured acceleration.

8 Constructing explanations

Table 3: F_{ramp} and acceleration prediction data

Calculated F_{ramp} (N)	Predicted acceleration (m/s²)	Measured acceleration (m/s²)	Net F_{ramp} (N)	Calculated - Net F_{ramp} (N)

a. Enter your calculated magnitude of F_{ramp} from Table 2. What direction was it in?

b. How did your measured acceleration compare to your predicted acceleration? What was the percent difference between them? Explain any differences.

c. If the accelerations were different, what does that mean about the magnitude of the net F_{ramp}?

d. Calculate the net F_{ramp} and record this in Table 3. How does it compare to your calculated F_{ramp}?

e. Find the difference in your results for F_{ramp} and record this value in the last column in Table 3.

f. Explain the difference you found between the two values for F_{ramp}. What is the percent difference? Would this difference be considered a force vector? If so, what is its magnitude and direction?

g. *Challenge:* Try this experiment with three marbles in the car. What do you find? Explain.

Investigation 5.2 Equilibrium of Forces and Hooke's Law

5.2 Equilibrium of Forces and Hooke's Law

How do you predict the force on a spring?

Springs come in two basic types. An extension spring is designed to be stretched, or *extended*. A compression spring is designed to be squeezed, or *compressed*. Hooke's law describes the relationship between force, the spring constant, and deformation.

In this investigation, you will:

- conduct experiments to determine the spring constants for an extension spring and a compression spring.
- create and test a graphical model for spring data.

Materials List
- Spring and Swings
- Electronic scale or triple beam balance
- 15 Washers

1 Experimenting with an extension spring

The goal of the first part of the investigation is to determine how much an extension spring extends per newton of applied force. You will measure the amount the spring extends as the applied force is changed. The data will allow you to determine the spring constant, which measures the strength of the spring.

1. Set up the Hooke's Law apparatus as shown at right.
2. Hang the blue-tabbed extension spring from the top of the stand, and attach the mass hanger with five washers on it.
3. Adjust the measurement card so the 0-centimeter mark lines up with the bottom of the mass hanger.
4. How much does the spring extend when you add more washers? Use Table 1 to help guide your investigation as you investigate the strength of the spring.

Table 1: Force and extension of a spring

Number of washers	Mass of holder and washers (kg)	Calculated force of holder and washers (N)	Spring's extension (cm)
5			0
7			
9			
11			

2 Creating a graphical model

a. Make a graph of force versus extension for the spring.

b. The force from a spring can be described by a formula known as Hooke's law. The spring constant (k) is a measure of the strength of the spring. For example, a spring with $k = 1$ N/cm produces 1 newton of force for every centimeter of extension.

HOOKE'S LAW (springs)
Force (N) $F = -kx$ — Spring constant (N/cm), Deformation (cm) (extension or compression)

Use your graph of force versus extension to determine the spring constant for the spring in the experiment. Express your result in N/cm.

3 Testing the model

a. Use your graph to predict how much the spring should extend for a hanger with 15 washers.

b. Set up and test the extension spring to see whether the graph gives the correct prediction.

c. How close did your prediction come to the actual extension of the spring? Calculate your percent error.

d. If you use the Hooke's law equation and the spring constant you found in 2b to predict the spring extension, you get a predicted extension that is greater than expected. Why is the answer off by that factor? Explain why you have to make a calculation adjustment to your experimental data to accommodate the Hooke's law equation prediction.

e. You have a second extension spring in the kit—this one has a white tab on it. Do you think this spring has a lower k value than the blue-tabbed spring you just studied, or a higher k value? Explain your choice.

f. Repeat Parts 1 and 2 of this investigation with the white-tabbed spring. How do the results compare to your prediction? Explain.

Investigation 5.2 Equilibrium of Forces and Hooke's Law

4 Experimenting with a compression spring

This part of the investigation uses a compression spring. Compression springs also obey Hooke's law, but they are designed to be compressed rather than extended. To measure the spring constant, you need to measure how much the spring compresses when different amounts of force are applied.

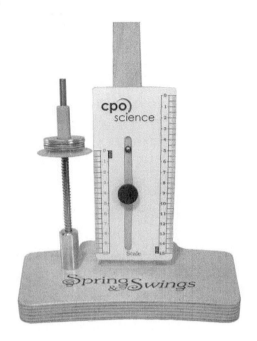

1. Place the compression spring on the spindle with the mass holder and three washers on top, as shown at right.
2. Adjust the measurement card so the 0-centimeter mark lines up with the bottom of the mass holder.
3. How much does the spring compress when you add more washers? Use Table 2 to help guide your experiment as you investigate the strength of the compression spring.

Table 2: Force and compression of a spring

Number of washers	Mass of holder and washers (kg)	Calculated force of holder and washers (N)	Spring's compression (cm)
3			0
6			
9			
12			

5 Creating a graphical model

a. Make a graph of force versus compression for the spring.

b. The force from a compression spring can also be described by Hooke's law. As with extension springs, the spring constant (k) is a measure of the strength of the spring. For example, a spring with $k = 1$ N/cm produces 1 newton of force for every centimeter of compression.

HOOKE'S LAW (springs)

$$F = -kx$$

Force (N), Spring constant (N/cm), Deformation (cm) (extension or compression)

Use your graph of force versus compression to determine the spring constant for the spring in the experiment. Express your result in N/cm.

6 Testing the model

a. Use your *graph* to predict how much the spring should compress for a holder with 15 washers.

b. Set up and test the compression spring to compare whether the graph gives the correct prediction.

c. How close did your prediction come to the actual compression of the spring? Calculate your percent error.

d. If you use the Hooke's law equation and the spring constant from 5b to predict the spring compression, you get a predicted compression that is higher than expected. Why is the answer off by that factor? Explain why you have to make a calculation adjustment to your experimental data to accommodate the Hooke's law equation prediction.

e. Could this spring be used to measure the mass of one washer? If so, describe a procedure for how to make the measurement. If not, explain why not.

f. How much force would it take to cause a compression of one millimeter?

5.3 Friction

What happens to the force of sliding friction as you add mass to a sled?

Friction is a resistive force that opposes motion. You encounter friction all the time. Without friction, cars wouldn't move, you couldn't write with a pencil, and you would have difficulty walking or sitting on a chair. In this investigation, you will explore the sliding friction between a sled and the SmartTrack.

In this investigation, you will:

- determine the force of friction present when you launch sleds of different mass on the SmartTrack.
- calculate the coefficient of sliding friction.

Materials List

- SmartTrack
- Energy Car sled
- Steel marbles
- Rubber band
- Track feet (2)
- Bubble level
- Velocity sensor
- DataCollector
- Electronic scale

1. Developing a model for sliding friction

1. Attach a foot to each end of the SmartTrack so it sits level on the table. Check it with the bubble level and adjust the feet as necessary to make the track level.
2. Put a rubber band onto the arms of the launcher in an *x* pattern by giving the rubber band a half-twist, then secure it in place between the retaining washers with the securing screws. Attach the launcher to the track facing away from the stopper end and adjust it so the rubber band is at the 100-centimeter mark, as shown in the picture above.
3. Place a sled with no marbles on the track and launch it.

a. What force causes the sled to move? What force opposes the sled's motion?
b. What happens to the sled's velocity as it travels along the track?
c. What would happen to the sled's motion if you added marbles to it? Why?

We want to study the force of friction that acts on the sled. It would be convenient if you could measure the force of sliding friction by dragging the sled along the track with a spring scale, but the friction force is too small to measure this way. Instead, you are going to use the velocity sensor to collect acceleration data for the sled's motion, and use Newton's second law to calculate the force of friction.

Think through this scenario: You launch the sled and it moves a certain distance along the track and comes to a stop. The net force that acts on the sled equals the force of friction. This friction force decelerates the sled. According to Newton's second law, the force equals the mass times the acceleration (or deceleration, in this case). So, if you know the sled's mass and deceleration (from the velocity sensor), you can calculate the force of friction that acts on the sled.

$$f = m \times a$$

force of friction = mass of sled × deceleration

2 Designing an investigation

a. How will adding mass to the sled affect the force of sliding friction? Develop a hypothesis to address this question. Your hypothesis should follow this format: "If the sled's mass affects the force of sliding friction, then _____."

b. *Predict*: When you carry out the investigation and see the velocity vs. time graph for the sled's motion, what will the graph look like?

c. *Predict*: When you carry out the investigation and see the acceleration vs. time graph for the sled's motion, what will the graph look like?

Design an investigation to test your hypothesis (2a). You will use the launcher to get the sled moving. Use the velocity sensor to measure the sled's acceleration, and use an electronic scale or triple beam balance to find the sled's mass. Vary the number of marbles. Here are some things to consider.

- Choose a launch force that will give a reasonable travel distance with all sled masses, and keep the launch force the same throughout the experiment.
- The sled decelerates, which is negative acceleration.
- Consider whether the force of friction is negative or positive.

d. Write out your investigation's procedure in numbered steps.

3 Arguing from evidence

a. Create a data table for all your measurements and calculations. Stay true to the units (kg, m/s², N).

b. Be sure to view velocity vs. time and acceleration vs. time graphs for each trial of the experiment. How do they compare to your predictions in 2b and 2c?

c. Was your hypothesis confirmed? Briefly summarize your results in a few sentences. Be sure to refer back to your hypothesis.

d. Use physics to describe why you got the results you did in this experiment.

e. The coefficient of friction (μ) is a ratio of the strength of sliding friction between two surfaces compared to the force (called the normal force) holding the surfaces together. In the case of the sled on the flat track, the normal force equals the weight of the sled. Write an equation for calculating the coefficient of friction between the sled and track.

f. Calculate the coefficient of sliding friction with all the friction force and weight data you collected. Compile the results in a data table, and find the average coefficient of sliding friction for your plastic-on-wood situation.

g. Draw a free-body diagram showing all the forces at work on the sled just as it is being launched.

Investigation　6.1　Launch Angle and Range

6.1 Launch Angle and Range

Which launch angle will give a marble the best range?

Any object moving through the air affected only by gravity is called a *projectile*. Projectile motion depends on the launch speed, launch angle, and the acceleration due to gravity. The combination of these factors creates a curved path called a *trajectory*. In this investigation, you will determine how a projectile's launch angle affects the horizontal distance it travels, called the *range*.

Materials List
- Marble Launcher with plastic marble
- Metric tape measure
- Calculator
- Masking tape
- Safety goggles

SAFETY
- **Never launch marbles at people.**
- **Wear safety goggles or other eye protection when launching marbles.**
- **Launch only the black plastic marbles that come with the marble launcher.**

1 Investigating projectile motion with the marble launcher

There are three experimental quantities that you can measure using the marble launcher.

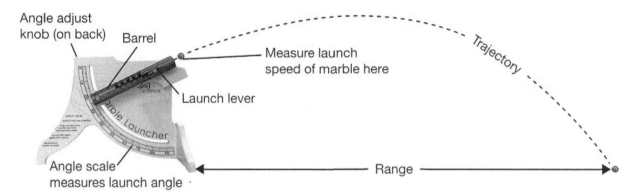

- **Launch speed** is the initial speed of the projectile. The launch speed for the marble launcher depends on the force applied by the spring inside the barrel. There are five notches that change the compression on the spring and give different launch speeds.

- **Launch angle**, also known as θ (theta), is the angle at which the projectile is launched. You can change the launch angle on the marble launcher by loosening the black knob on the back and adjusting the angle of the barrel between 0 and 90 degrees.

- **Range** is the horizontal distance that the projectile travels before touching down. The range of the projectile is dependent on the launch speed and the launch angle.

a. How do you think *launch speed* affects the range of the marble?
b. How do you think *launch angle* affects the range of the marble?

Launch Angle and Range   Investigation   6.1

2 Setting up

For this experiment, you will investigate the relationship between launch angle and range.

1. Follow all safety rules.
2. Use the fifth slot of the barrel to launch the marble for each trial. Pull the launch lever back and slip it sideways into the fifth slot. Put a marble in the end of the barrel. The marble launcher is now ready to launch.

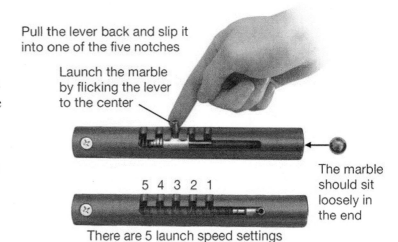

3. You will change the angle for each launch starting at 10 degrees and increasing five degrees up to 85 degrees.
4. A minimum of two people are needed per launcher. One person releases the launch lever and the other watches where the marble lands. A few launches should be done at each angle to be sure that the data is accurate. It also takes a few times to accurately find the spot where the marble lands.
5. Use a strip of masking tape on the floor to make sure that the marble launcher is set back in the same place every time. A tape measure along the floor provides a good way to measure distance.

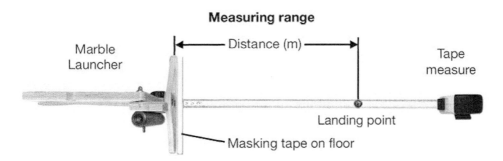

Investigation 6.1 Launch Angle and Range

3 Doing the experiment

1. Use notch 5 for each launch.
2. Conduct three trials for each launch angle, and find the average range for each angle. Record your data in the table.

Launch angle (degrees)	Trial 1 range (meters)	Trial 2 range (meters)	Trial 3 range (meters)	Average range (meters)
10				
15				
20				
25				
30				
35				
40				
45				
50				
55				
60				
65				
70				
75				
80				
85				

4 Creating a graphical model

a. Plot a graph of the average range (*y*-axis) versus the launch angle (*x*-axis).

b. After looking at your graph, specify which angle gives the maximum range.

c. Why is the range less for angles larger or smaller than the angle corresponding to the maximum range?

5 Constructing explanations

a. Look at your data. How close to your predicted location can you expect your marble to land? Write your answer in this format: The marble consistently lands within ± _____ centimeters of my predicted location.

b. Suppose the marble launcher is set to 62 degrees. Using your graph, predict how far the marble will go. Express your prediction as a "plus-or-minus" range.

c. Express the error in your prediction as a percentage.

d. Launch a marble at 62 degrees and measure the range. How accurate is your prediction? Calculate your percent error.

e. How could you make your prediction more accurate?

f. Three marbles are launched using the third notch on the barrel. Each launch has a different angle. Which launch will have the greatest range?

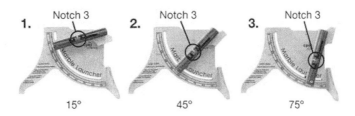

6.2 Launch Speed and Range

How does launch speed affect the range of a projectile?

In the last investigation, you identified how the launch angle of a projectile affects its *range*. Through your experiments, you determined that there is a certain launch angle that allows for the maximum range of the marble. In this investigation, you will explore the relationship between launch speed and range. The different settings for the spring on the marble launcher will allow you to get five different launch speeds. The DataCollector and the photogate will provide accurate speed measurements that will help you identify the relationship.

Materials List
- Marble Launcher
- DataCollector
- Photogate
- Plastic marble
- Metric tape measure
- Calculator
- Safety goggles
- Masking tape

SAFETY
- **Never launch marbles at people.**
- **Wear safety goggles or other eye protection when launching marbles.**
- **Launch only the black plastic marbles that come with the marble launcher.**

1 Setting up the experiment

1. Just as in the last experiment, you will need to make a tape mark on the floor and use a tape measure to measure the range of the marble.
2. For this experiment, keep the angle constant (45 degrees) and change the launch speed as shown.
3. Put the DataCollector in Timer mode and use the Interval function with one photogate connected. If you attach the photogate correctly, the center of the marble crosses the light beam.
4. The speed of the marble is calculated by dividing the diameter of the marble (0.019 meters) by the time that the beam is broken (time from photogate A).

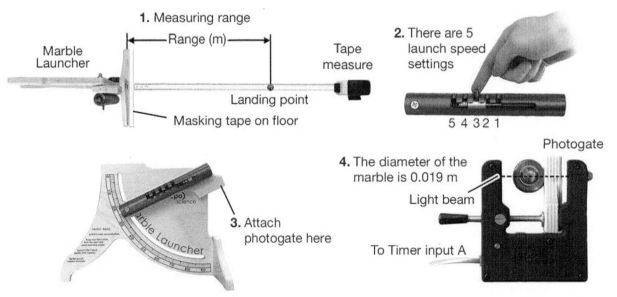

Launch Speed and Range   Investigation   6.2

2 Doing the experiment

1. For this experiment, you will vary the launch speed by changing the spring setting.
2. Use a fixed launch angle (an angle near 45 degrees works best) for the entire experiment. Record your launch angle here: _____.
3. Launch the marble three times at each spring setting. Record the time from photogate A and the range for each trial in the table below.
4. Find the average time and average range for the three trials. Record these values in the table.
5. Calculate the launch speed by dividing width of the marble by the average time.
6. The shaded columns of the table contain the data that you will be comparing (launch speed and average range).

Spring setting (1 to 5)	Width of marble (m)	Time from photogate A (s)			Average time (s)	Launch speed (m/s)	Range (m)			Average range (m)
1	0.019									
2	0.019									
3	0.019									
4	0.019									
5	0.019									

3 Arguing from evidence

a. Look at your data. Do you see a relationship between the launch speed and the average range?
b. Make a graph showing the average range of the marble (*y*-axis) vs. launch speed (*x*-axis).
c. Describe the graph. Is the graph a straight line or a curve?
d. How is this graph different from the range vs. angle graph you made in the last investigation?
e. Can you determine an *exact* mathematical relationship between average range and launch speed from the graph? Why or why not?

Investigation 6.3 Levers and Rotational Equilibrium

6.3 Levers and Rotational Equilibrium

How do levers work?

The development of the technology that created computers, cars, and the space shuttle began with the invention of *simple machines*. A simple machine is a mechanical device that does not have a source of power and accomplishes a task with only one movement. The *lever* is an example of a simple machine. A lever allows you to move a rock that weighs 10 times as much as you do (or more)!

Materials List
- Physics Stand
- Levers set
- Set of weights
- Spring scales capable of measuring forces from 0 to 10 N
- 20 Glass marbles (from Atom Building Game)

1 How do levers work?

A lever includes a stiff structure (the *lever*) that rotates around a fixed point called the *fulcrum*. The side of the lever where the input force is applied is called the *input arm*. The *output arm* is the end of the lever that moves the rock or lifts the heavy weight.

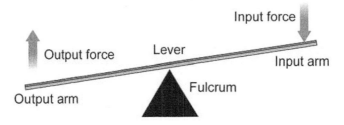

Levers use *torque* to lift or move objects. Torque is a force applied over a distance that causes rotation to occur. In a lever, the force is applied *perpendicular* to the distance. This perpendicular force causes rotation to occur. To calculate torque, you multiply the force applied by the distance of the force from the fulcrum. Torque is measured in units of force (newtons) times distance (meters), or *newton-meters* (N-m). The example below shows how to calculate torque.

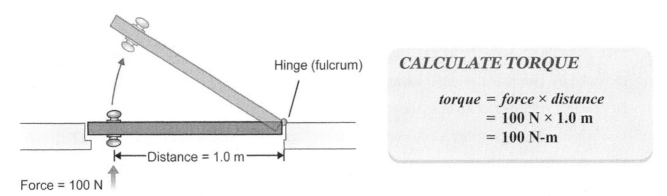

CALCULATE TORQUE

$$torque = force \times distance$$
$$= 100 \text{ N} \times 1.0 \text{ m}$$
$$= 100 \text{ N-m}$$

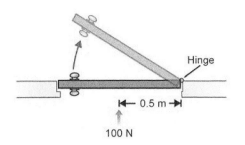

Use the graphic to the left to answer the following questions.

a. Calculate torque in the example.

b. Based on your answer, why is a doorknob placed as far away from the hinge as possible?

c. What would happen if you applied the same amount of force directly to the hinge? (*Hint*: Find the torque.)

2 Levers and torque

1. Set up the lever as shown in the diagram below. For this part of the investigation, you can secure the mass to the lever by looping the string around the distance-indicating post on the lever as shown in the diagram below. This way, you will know the exact distance of the mass from the fulcrum.

2. Use a spring scale to measure the force exerted by a weight.

3. Calculate the *input torque* by multiplying the amount of force on the input arm by the distance from the fulcrum.

4. Calculate the *output torque* by multiplying the amount of force on the output arm by the distance from the fulcrum. Show this as a negative value because it causes the lever to rotate in the *opposite* direction from the input torque's.

5. Conduct more trials by hanging different combinations of weights on both sides of the fulcrum so that the lever balances. Record your results in the chart on the next page. There is room for six trials.

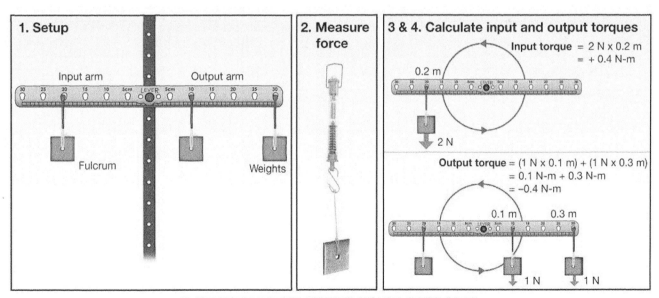

Investigation 6.3 Levers and Rotational Equilibrium

Data Chart

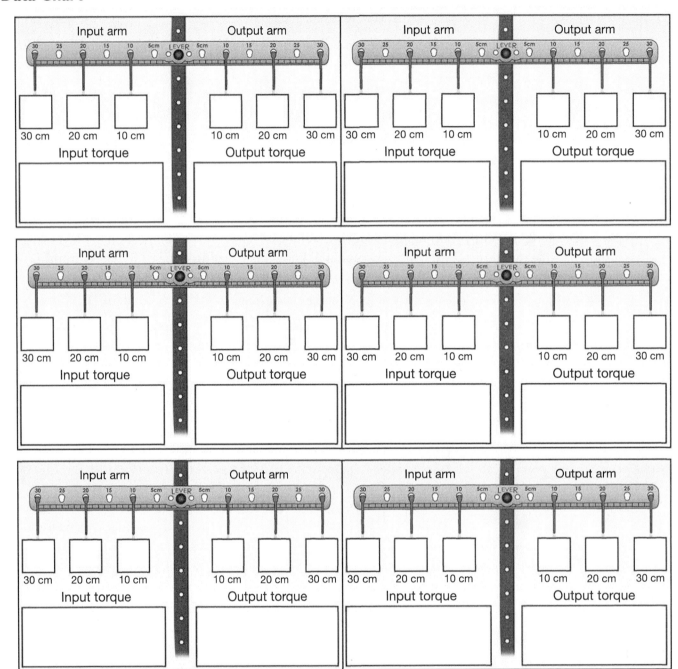

3 Constructing explanations

a. How do the input and output torques compare for each trial?

b. Explain, using your knowledge of torque, why the lever is in balance.

c. Use your data to identify the mathematical relationship between input force, input length, output force, and output length. Write the relationship as a mathematical formula.

d. Why do you think the term *rotational equilibrium* is used to describe when the lever balances?

4 Using the lever to solve a mystery

You have seen what it takes to make the lever reach equilibrium and be in balance. Now it is up to you to use what you have learned about the mathematical relationship between both sides of the lever to uncover the mass of one single glass marble. You may consider using the mass hanging clips for this part. You can move hanging masses to any location on the lever using these, and they will clearly indicate their exact distance from the fulcrum.

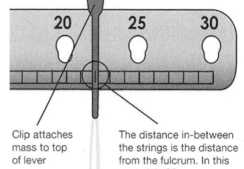

Clip attaches mass to top of lever

The distance in-between the strings is the distance from the fulcrum. In this case, it is 22 cm.

a. What condition had to be in place for the lever to balance?

b. You have 20 glass marbles and a small bag that can be used to hang them from the lever. (1) Describe a process that you could follow using the lever and weights to determine the mass of one single marble. (2) Explain what each step would be and why it is being done, and be sure to include terms like *input torque, output torque, equilibrium, force,* and *distance*. (3) The mass of one single marble must be given in grams, so include how you would determine that from what you know about the relationship between force and mass.

c. Once you have completed your plan, use it to determine the mass of one single marble using the lever and weights. What was your result?

d. How did your result compare to that of the other groups in your class?

e. Use an electronic scale or a triple beam balance to determine the mass of one single marble. How does this compare to your result?

f. Is getting the mass of one single marble the best way to evaluate your result? What might be a better method? (*Hint*: Do you think all of the marbles have the exact same mass?)

g. Use your method and compare your result to this new figure. Are you satisfied with your result?

h. If time permits, discuss with your class all of the things you could have done to refine your experiment. Then try these methods, and see if you can get a closer result. Was your result much closer?

Investigation 7.1 Force, Work, and Machines

7.1 Force, Work, and Machines

How do simple machines affect work?

Machines are things humans invent to make tasks easier. Simple machines work by using directly-applied forces. Simple machines allowed humans to build the great pyramids and other monuments using only muscle power. This investigation is about how simple machines use force to accomplish a task.

Materials List
- Ropes and Pulleys
- 4 Steel weights
- Meter stick or tape measure
- Spring scales
- Physics Stand

1 Building a simple machine

1. Attach four weights to the bottom block. Use a spring scale to measure the weight of the bottom block and record it as the output force.
2. Attach the top block near the top of the physics stand.
3. Thread the yellow string over one or more of the pulleys of the top and bottom pulley blocks. The yellow string can be clipped to either the top block or the bottom block.
4. Build combinations with 1, 2, 3, 4, 5, and 6 supporting strings directly supporting the bottom block. (*Hint*: 1, 3, and 5 have the string clipped to the bottom block; 2, 4, and 6 have the string clipped to the top block.)
5. Use a force scale to measure the force needed to slowly lift the bottom block for different combinations of supporting strings.

Safety Tip: Don't pull sideways, or you can tip the stand over.

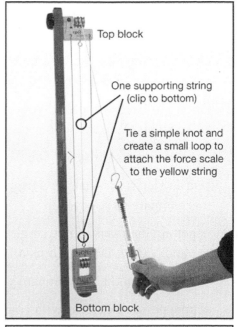

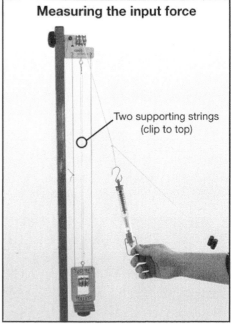

Measuring the input force

54

Table 1: Input and output forces

Number of supporting strings	Input force (newtons)	Output force (newtons)
1		
2		
3		
4		
5		
6		

2 Constructing explanations

a. As you increase the number of supporting strings, what happens to the force needed to lift the bottom block?

b. Write a rule that relates the number of supporting strings, input force, and output force.

c. Research the term *mechanical advantage*. What does this mean for a simple machine?

d. Use your data from Table 1 to calculate the mechanical advantage for each arrangement of supporting strings.

3 The input and output distance

1. Use the marker stop (cord stop) to mark where the string leaves the top pulley.
2. Choose a distance that you will lift the bottom pulley during each trial of the experiment. This is the *output distance*. Your output distance should be at least 20 centimeters.
3. Pull the yellow string to lift the block your chosen distance.
4. Measure how much string length you had to pull to lift the block. This is the *input distance*.
5. Measure the input and output distances for each of the different configurations (1, 2, 3, 4, 5, and 6).
6. Copy your input force and output force data from Part 1 into Table 2.

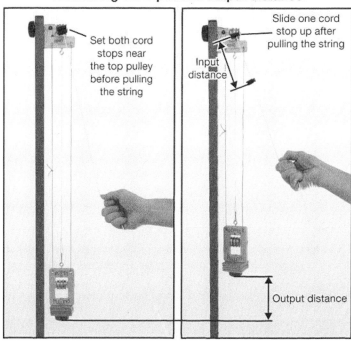

Measuring the input and output distance

Investigation 7.1 Force, Work, and Machines

Table 2: Force and distance data

Mechanical advantage	Output force (newtons)	Output distance (meters)	Input force (newtons)	Input distance (meters)
1				
2				
3				
4				
5				
6				

4 Arguing from evidence

a. As the mechanical advantage increases, what happens to the length of the string you have to pull to raise the block?

b. The word *work* is used in many different ways. For example, you *work* on science problems, your toaster doesn't *work*, or taking out the trash is too much *work*. In science, however, *work* has one specific meaning. Write one sentence that defines work in its scientific meaning.

c. You may have heard the saying, "nothing is free." Explain why this is true of the ropes and pulleys. (*Hint*: What do you trade for using less input force to lift the block?)

d. Use your data to calculate the work done on the block (the *output work*).

e. Next, use your data to calculate the work you did as you pulled on the string to lift the block. This is the *input work*.

Table 3: Output and input work

Mechanical advantage	Output work (joules)	Input work (joules)
1		
2		
3		
4		
5		
6		

f. For each mechanical advantage, how do output and input work compare?

g. Is output work ever greater than input work? Can you explain this?

h. Explain any differences between input and output work in your data.

7.2 Work and Energy

How does a system get energy?

Energy comes from somewhere. When you lift a box off the floor, the increase in energy of the box comes from the work you do on the box. This investigation looks into the conversion of work to energy.

Materials List
- Mass balance
- 10-N Force scale
- Energy Car
- Physics Stand
- DataCollector
- Metric ruler
- String
- SmartTrack
- 3 Steel marbles
- Photogate
- Rubber band

1 Measuring the work done

Thread a knotted string through the hole in the car

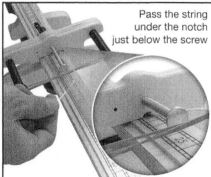
Pass the string under the notch just below the screw

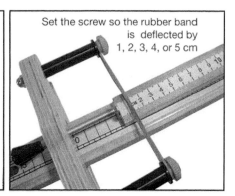
Set the screw so the rubber band is deflected by 1, 2, 3, 4, or 5 cm

Use a force scale to measure the force it takes to pull the car so it just touches the screw.

1. Attach one foot to each end of the SmartTrack and set it up horizontally on your lab table.
2. Attach the launcher to the SmartTrack at the 20-centimeter mark pointing toward the catcher.
3. Put a rubber band onto the arms of the launcher in an *x* pattern by giving the rubber band a half-twist, then secure it in place between the retaining washers with the securing screws.
4. Adjust the threaded screw until the distance between the screw and the front of the rubber band is one centimeter (see diagram).
5. Tie a knot in the string and pass the string through the hole in the car and under the notch just below the screw. Tie a loop in the other end of the string.
6. Use a spring scale to measure the force when the car is just touching the screw.
7. Adjust the screw for distances of 2, 3, 4, and 5 centimeters. Measure the force for each distance. Record your measurements in Table 1.

Table 1: Force vs. distance data

Distance rubber band is stretched (cm)	Force (N)

2 Creating a graphical model

a. Graph the force from the rubber band versus the distance.

b. Write a one-sentence definition of work in physics.

c. You would like to know how much work the rubber band does on the car during a launch. Since the force changes with distance, use the graph to do some averaging. Divide your graph up into bars, each representing one centimeter of distance. Make the height of each bar the average force over the distance interval covered by the bar. The area of each bar is the work done over that interval of distance. Your graph will have data out to five centimeters. The sample graph below shows data from zero to three centimeters as a demonstration that you can follow for your entire graph.

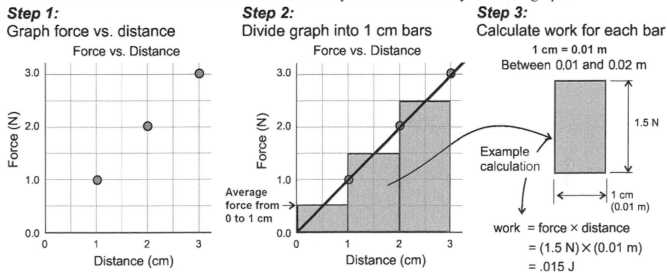

d. To get the total work done on the car, add up all the work done as the rubber band straightens out and moves the car forward. Use the table below to calculate the work done.

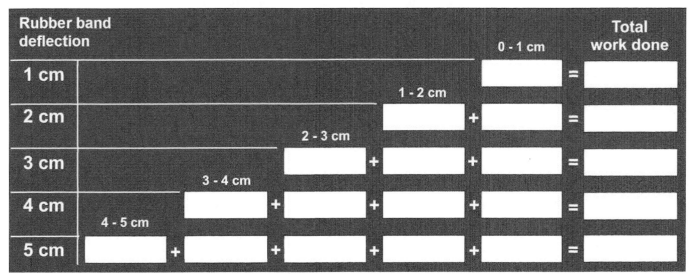

e. Make a graph showing the work done on the y-axis and the deflection of the rubber band on the x-axis.

f. Assume all the work done becomes kinetic energy of the car. Derive a formula for the speed of the car that depends only on the car's mass and the work done by the rubber band.

Investigation 7.2 Work and Energy

3 Testing the theory

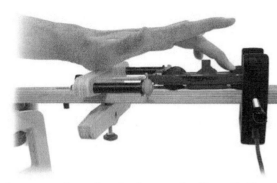

Set the photogate just ahead of the flag when the rubber band is straight.

Launch the car at the same measured deflections for which you measured the force.

1. Plug a photogate into the A slot on the DataCollector. Select the timer mode of the DataCollector and use the interval function.
2. Put the photogate on the track so the flag on the car breaks the light beam about one centimeter after leaving the rubber band. Your teacher will assign you a number of steel marbles to add to the car. Find the mass of the car after adding the marbles. Keep this number of marbles in the car throughout the procedure.
3. Use the adjustment screw to launch the car at the same measured deflections of the rubber band for which you measured the forces (1, 2, 3, 4, and 5 centimeters).
4. Record the time taken for the car to move through the photogate in Table 2. When releasing the car during the launch, be sure not to break the photogate's beam with your fingers.
5. At each deflection, take data with the same number of steel marbles in the car each time. Use the time through photogate A to calculate the speed of the car in the column for "measured speed" in Table 2.

Table 2: Deflection, mass, and speed data

Deflection of rubber band (cm)	Mass of car (kg)	Photogate time (s)	Measured speed (m/s)	Predicted speed (m/s)

4 Arguing from evidence

a. Use your formula to predict the speed the car should have at each combination of mass and deflection. Write the results in Table 2 in the column "Predicted speed."

b. Graph the predicted speed of the car versus the deflection of the rubber band. Draw a smooth curve through the plotted points. On the same graph, show the measured speeds.

c. Graph the measured speed of the car versus the deflection of the rubber band on the same graph as part c.

d. Does your data support the theory that the energy of the car is equal to the work done by the rubber band? Your answer should provide evidence from your results and discuss possible sources of errors.

e. *Challenge:* Use your theory to predict the speed if cars with a different number of steel marbles are launched at a deflection of three centimeters. Do the experiment and see if your prediction is accurate.

7.3 Energy and Efficiency

How well is energy transformed from one form to another?

According Newton's laws, you could start a car moving in a frictionless world and it would continue with the same speed forever. The real world is never frictionless, however, so the car slows down. In fact, all real processes that exchange energy lose small amounts to heat and the wearing away of material due to friction. This investigation is about efficiency, the property that describes how well energy is transformed in a process.

Materials List
- Energy car
- SmartTrack
- DataCollector
- Meter stick
- Photogate
- Steel marbles
- Bubble level
- Mass balance
- Graph paper

1 Kinetic energy exchange

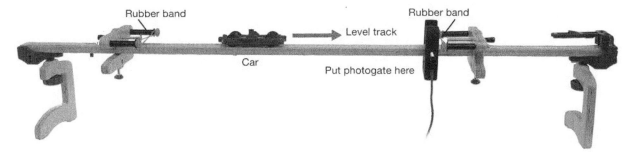

1. Set up the SmartTrack with two feet on your lab table so it is level. Place two launchers on it so they are facing each other; one at the 20-centimeter mark and one at the 90-centimeter mark. You will need to share a launcher with another group for this activity.
2. Put a rubber band onto the arms of both launchers in an *x* pattern by giving the rubber band a half-twist, then secure it in place between the retaining washers with the securing screws.
3. Connect a photogate to the A slot on the DataCollector. Select Timer mode and use the Interval function to measure the time the car takes to move through photogate A.
4. Position the photogate so the flag breaks the light beam just before hitting the rubber band on the launcher at the 90-centimeter mark.
5. Use the launcher at the 20-centimeter mark to launch the car down the track so it goes through the photogate, bounces off the other launcher's rubber band, and travels back through the photogate. Stop the car after it moves back through the photogate the second time.
6. Measure the time through the photogate before and after the car bounces off the rubber band. You will need to use the memory function to display the "before" time. Record the time in Table 1.
7. Record at least two trials with consistent data and calculate the average speeds before and after hitting the rubber bands.
8. Calculate the velocity of the car before and after it bounces off the rubber band.
9. Measure the mass of the car and do the experiment for several different masses. Record your data in Table 1.

Investigation 7.3 Energy and Efficiency

Table 1: Kinetic energy data

Mass of the car (kg)	Time before collision (s)	Velocity before collision (m/s)	Time after collision (s)	Velocity after collision (m/s)

2 Using your model

a. Describe the energy flows that occur between the car heading toward the rubber band and the car leaving the rubber band.

b. If the transformation of energy were perfect (100 percent efficient) what would you expect the speed of the car to be before and after the collision with the rubber band?

c. Write down the formula for kinetic energy and use the formula to calculate the kinetic energy of the car before and after bouncing off the rubber band.

d. Calculate the efficiency of the process of bouncing the car off of a rubber band.

Useful relationships

$$\text{Kinetic energy} = \frac{1}{2}mv^2$$

$$\text{Efficiency} = \frac{\text{Final energy}}{\text{Initial energy}}$$

3 How does the efficiency change?

1. Adjust the deflection of the rubber band you use to launch the car to get different speeds when you launch it. Bounce the car off the other rubber band at different velocities and record the photogate's data in Table 2.

Table 2: Energy efficiency data

Mass of the car (kg)	Time before collision (s)	Velocity before collision (m/s)	Time after collision (s)	Velocity after collision (m/s)

4 Constructing explanations

a. Calculate the efficiency of the rubber band for the different masses and velocities you tested.

b. Plot a graph showing the efficiency on the vertical (*y*) axis and the velocity on the horizontal (*x*) axis. How does the efficiency change with the speed of the car?

c. Try changing the tension of the rubber band. Does this affect the efficiency?

8.1 Energy Flow in a System

How does the energy move through a series of transformations?

Most systems in the world exchange energy in many forms as they operate. To understand how things work, we often trace the flow of energy. The performance of a machine can be improved by working at improving each step where energy is transformed. This Investigation traces the flow of energy in a three-step process using the Energy Car.

Materials List

- Energy Car
- SmartTrack
- DataCollector
- Physics Stand
- Meter stick
- Photogate
- Electronic scale
- Steel marbles (3)

1 Tracing the energy through the system

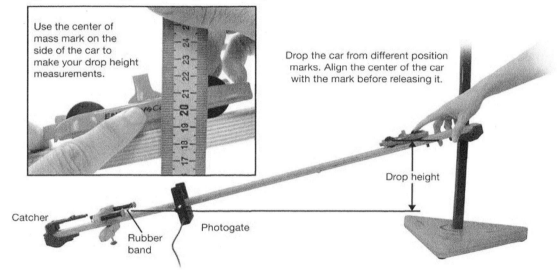

1. Attach the SmartTrack to the fifth hole up from the bottom of the physics stand.
2. Adjust the plunger screw of the launcher so it is extended all the way back and in contact with the face of the launcher. Place the launcher on the SmartTrack immediately after the 100-centimeter mark.
3. Put a rubber band onto the arms of the launcher in an *x* pattern by giving the rubber band a half-twist, then secure it in place between the retaining washers with the securing screws.
4. Put a photogate at the 80-centimeter mark of the SmartTrack. Connect it to the A slot on the back of the DataCollector.
5. Use the DataCollector in Timer mode and the Interval function with the photogate to measure the speed of the car before and after it bounces off the rubber band. The Memory feature can be used to see the time the car's flag breaks the beam of the photogate before and after the bounce.
6. Measure the drop height of the car in reference to its height when it hits the rubber band against the launcher's plunger screw. Use the center-of-mass mark on the side of the car to make your drop height measurements. Record your measurements in Table 1.
7. Drop the car from several heights. Calculate and record the car's speed for each trial in Table 1.
8. Measure and record the mass of the car in Table 1. Do the experiment for several different masses.

Investigation 8.1 Energy Flow in a System

Table 1: Energy data

Drop Height (m)	Mass of car (kg)	Time before rubber band (s)	Speed before rubber band (m/s)	Time after rubber band (s)	Speed after rubber band (m/s)

2 Constructing explanations

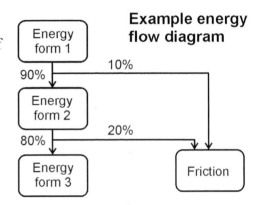

Example energy flow diagram

a. What three forms of energy are most important to the motion of the car in this system?

b. Calculate the total energy of the car in joules at three places:
   (1) At the top of the hill before it is dropped.
   (2) At the photogate heading into the rubber band for the first time.
   (3) At the photogate after bouncing off the rubber band.

c. Describe the three most important energy transformations that occur during the motion (other than friction).

d. Draw an energy flow diagram showing the measured and calculated amounts of energy at each of the three measured places. Label any energy that is lost as "friction."

e. What percentage of its initial energy does the car have after passing through the photogate for the second time? Assume the car has "lost" any energy spent overcoming friction.

f. Where does the energy "lost" to friction go? Is the energy really destroyed?

3 Improving the overall performance

a. Suggest a modification you can make to the system that would leave the car with a higher percentage of energy after its second pass through the photogate.

b. Explain why you believe your modification will result in higher energy efficiency.

c. Write a few sentences describing a procedure to test your modification.

4 Planning your investigation

a. Plan an investigation that will test your idea for improving overall energy efficiency. Write a procedure for doing the investigation. Identify which data you expect to record and how the data will allow you to evaluate your idea.

b. Set up and do the investigation you designed.

c. Analyze your results. Compare the percentage of energy the car has after the second pass through the photogate to what it was in the earlier experiments. Your answer must use data you collected.

d. Give at least one reason why the efficiency is higher, lower, or about the same, compared to what it was in the investigation.

8.2 People Power

What are your work and power as you climb a flight of stairs?

When you walk up a flight of stairs, you do work to lift your body against the force of gravity. If you know your weight and the vertical distance you climb, you can calculate the work you do. Measuring the time it takes to climb the stairs allows you also to calculate your power. In this investigation, you will calculate, then compare, the work and power of several students in your class as they climb a set of stairs.

Materials List
- DataCollector or simple stopwatch
- Meter stick
- Bathroom scale

1 Collecting the data

1. You will be doing this activity as a class. Choose several students to volunteer to climb a flight of stairs as other students measure the time.

2. Use a bathroom scale to measure the weight of each volunteer. Convert each weight to newtons using the conversion factor: 1 pound = 4.448 newtons. Record each person's weight and name in Table 1.

Table 1: Stair climbing data

Name	Weight (N)	Stair height (m)	Time 1 (s)	Time 2 (s)	Time 3 (s)	Average time (s)

3. Choose three students to be timekeepers and two students to be stair-height measurers. These should not be the same students who are climbing the stairs.

4. Locate an empty stairwell with one or two flights of stairs. The stair-height measurers should measure the total vertical distance a person climbs when going from the top to the bottom. The easiest way to do this is to measure the height of one stair, count the number of stairs, and multiply.

5. Have each person listed in Table 1 climb the stairs once as the three timekeepers measure the time with the DataCollector in stopwatch mode or a simple stopwatch. Record all of the times in Table 1. Calculate the average time for each person.

Investigation 8.2 People Power

2 Calculating work and power

1. Calculate the work done by each person. The force for each person is his or her weight. Record the work in Table 2.
2. Use each person's average time to calculate the power. Record each power in Table 2.

Table 2: Work and power

Name	Work (J)	Power (W)

3 Analyzing the data

a. Who did the most work? What do you notice about this person's weight?
b. Who did the least work? What do you notice about this person's weight?
c. Who had the greatest power? Must this be the person with the fastest time?
d. Who had the least power? Must this be the person with the slowest time?
e. Calculate the average work for the students.
f. The Calorie is also a unit of work. One Calorie equals 4,186 joules. Calculate the average number of Calories of work done by the students. You may be surprised at how small the answer is!
g. Calculate the average power of the students.
h. A typical bright light bulb has a power of 100 watts. How does this compare to the average power of the students?
i. Imagine that two people of equal weights climb the same flight of stairs. One runs, and the other walks. Do they burn the same number of Calories? Do they have the same power? Explain.
j. Juan's weight is twice the weight of his little sister, Anna. They climb the same set of stairs and find that they have the same power. Explain how this can be possible.

8.3 Transportation Efficiency

Which transportation method is the most efficient?

When deciding how to get from one place to another, we usually choose our method of transportation based on convenience. If you are traveling within a city, you may take the subway or walk. When going from Boston to Los Angeles, you would probably fly. All forms of transportation require energy. In this investigation, you will determine which method is the most efficient.

Materials List
- Computer with internet connection
- Calculator

1 Thinking about efficiency

The efficiency of a process is found by dividing the output work (or energy) by the input work. The efficiency of a car's engine could be calculated by dividing the work done to give the car kinetic and potential energy by the energy contained in the gasoline. Multiplying by 100 converts the efficiency into a percent.

Suppose you want to calculate the efficiencies of a small car and a large SUV as each drives the same distance between your house and the grocery store. The car uses an amount of gasoline that contains 2,000,000 J (2 megajoules, or MJ) of energy, and the SUV uses gasoline with 6,000,000 J of energy. The car does 360,000 J of work to move itself and two passengers, while the SUV does 1,140,000 J of work.

a. Which vehicle uses more energy? Which does more work?

b. Calculate the efficiency of each vehicle. Which is more efficient?

c. Why does the SUV do more work while traveling the same distance?

d. Do you think efficiency is a good way to compare vehicles to determine which is best for the environment? Why?

2 Comparing vehicles

To better compare vehicles, fuel consumption is calculated in miles per gallon (mpg). If you go to a car dealer, you will see stickers on all of the cars that state the mpg in the city and on the highway. These are averages, as the value depends on many factors such as traffic conditions, terrain, and the speed of the car. At higher speeds, an increase in air resistance reduces the efficiency.

When comparing cars with other methods of transportation, such as buses, planes, and trains, the number of passengers must be considered. The efficiency of a bus getting five mpg but carrying 50 passengers is equivalent to a car carrying one passenger and getting 250 mpg.

Not all vehicles use gasoline. Subway trains use electricity supplied through the rails or overhead wires. Plug-in electric cars use electricity to charge the batteries that power the cars. Other cars use natural gas. To make comparisons between these vehicles and gasoline-powered ones, their manufacturers calculate a mpg equivalent.

Investigation 8.3 Transportation Efficiency

a. Research online to find the fuel consumption for at least five different types of vehicles. Record your results in Table 1. Include the number of passengers each vehicle typically carries.

Table 1: Vehicle fuel consumption

Type of vehicle	Mpg	Average number of passengers	Adjusted mpg

b. Calculate the adjusted mpg by multiplying the mpg by the average number of passengers.

c. Discuss what your research shows you about the fuel consumption of different vehicles.

3 Human-powered transportation

Gasoline, natural gas, and the fuel used at most power plants are all fossil fuels. These forms of energy are nonrenewable. Once they are used up, we will not be able to produce more. Walking and biking use energy we get from food, a renewable source of energy. These methods of transportation do not produce the air pollution that is created when fossil fuels are burned, and they are healthy forms of exercise.

Energy used for human transportation is calculated in food calories per mile. A food calorie (also called a Calorie or kilocalorie) is equal to 4,184 joules.

a. Do research to find the energy used per mile of biking or walking. If you have time, you can also include activities such as running, inline skating, kayaking, and swimming. Record your data in Table 2. Make sure you find values for calories per mile, not calories per hour.

Table 2: Energy use in human-powered transportation

Activity	Food calories per mile

b. When doing your research, what factors did you find affected the calories burned for a mile of biking or walking? Why do these factors matter?

4 Comparing vehicle and human-powered transportation

A gallon of gasoline contains approximately 1.3×10^8 J (130 MJ) of energy, so fuel consumption in mpg can be converted into miles per joule or joules per mile.

a. Calculate the number of joules required to bike one mile and walk one mile.

b. Calculate the number of joules of energy used to travel one mile for each of the vehicles listed in Table 1.

c. Which method of transportation requires the least energy? Discuss the reasons why you think so.

9.1 Temperature and Heat

How is temperature different from energy?

Hot and cold are familiar sensations, but you may not have known that they are caused by the energy of atoms! This investigation will explore the concepts of temperature, and the difference between temperature and thermal energy (heat).

Materials List
- DataCollector
- Temperature probe
- 10 Steel washers
- 2 Foam cups
- Calculator
- Permanent marker
- Ice
- Cold water
- Hot water
- Electronic scale or triple beam balance

 Use caution when handling hot water.

1 Thinking about temperature and energy

Consider the following experiment. Two foam cups contain equal masses of water. One cup contains cold water with a temperature of 0°C. The other contains hot water with a temperature of 50°C. The hot water is mixed with the cold water and stirred.

a. Which cup has more energy, the hot one or the cold one? Why do you think so?

b. What do you think the temperature of the mixture will be? Why?

c. If the system includes both the cold and hot water, compare the energy of the system before mixing to its energy after mixing. You may ignore any energy going to air or friction.

2 Doing the experiment

1. Prepare foam cups containing 100 grams each of hot and cold water.
2. Measure and record the temperatures before mixing.
3. Mix the water, stir well, and measure the final temperature.

Table 1: Temperature data for mixing equal masses of water

Cold water temperature before mixing (°C)	Hot water temperature before mixing (°C)	Mixture temperature (°C)

Investigation 9.1 Temperature and Heat

3 Constructing explanations

a. Given the starting hot and cold temperatures, what do you think the mixture's temperature should be?
b. Did the result of the experiment agree with your prediction? Discuss the meaning of *agree* in terms of the accuracy of your experiment.

4 Doing an experiment with metal and water

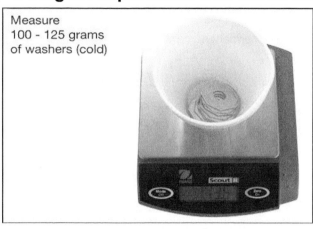

Measure 100 - 125 grams of washers (cold)

Add the same mass of hot water to the cold washers

What is the mixture temperature?

1. Put enough washers in a foam cup so the mass is between 150 and 200 grams. Record the mass of the washers.
2. Cover the washers with ice and water so they become cold.
3. Prepare a mass of hot water equal to the mass of the washers in another foam cup. Measure its temperature and record in Table 2.
4. Record the temperature of the cold water and washers, then pour off all the water leaving just the washers in the cup.
5. Add the equal mass of hot water to the cup with the washers.
6. Mix the water, stir well, and measure the final temperature.

Table 2: Temperature data for combining water and steel washers

Washer mass (kg)	Washer temperature before mixing (°C)	Hot water mass (kg)	Hot water temperature before mixing (°C)	Mixture temperature (°C)

5 Constructing explanations

a. Why didn't the temperature of the steel and water mixture come out halfway between the cold temperature of the washers and the hot temperature of the water, even though you mixed equal masses?
b. Different materials have different abilities to store thermal energy. Research and describe the property of a material that measures its ability to store thermal energy. What units does this property have?
c. How much energy does it take to raise the temperature of a kilogram of steel by 1°C?
d. *Challenge:* Suppose you drop 0.5 kg of steel at 100.0°C into a bucket containing 2.0 kg of water at 0.0°C. What is the final temperature of the mixture? (*Hint*: Apply energy conservation.)

9.2 Energy and Phase Changes

How is energy involved when matter changes phase?

We experience matter in three phases: solid, liquid, and gas. Changing from one phase to another means changing the bonds between atoms. When this happens, energy must be either absorbed or given off. This investigation will explore how much energy it takes to change matter from one phase to the next.

Materials List
- DataCollector
- Temperature probe
- 4 Foam cups
- Permanent marker
- Ice
- Cold water
- Hot water
- Electronic scale or triple beam balance

 Use caution when handling hot water.

1 Doing the experiment

Measure equal masses into all four cups (113.6 g is an example only, use your own mass)

1. Place some crushed ice in cold water, then transfer at least 100 grams of ice into a cup. Try not to get any liquid water, just ice.
2. Measure the mass of the ice and cup.
3. Prepare another cup with an equal mass of ice-cold water (with ice removed).
4. Prepare two more cups with equal masses of hot water.
5. Measure and record the temperatures of all four cups before mixing. Assume the solid ice is at 0°C.
6. Mix the ice and hot water in one cup and the hot and cold water in another cup. Stir well, and measure the final temperature of each mixture after all the ice has melted.

71

Investigation 9.2 Energy and Phase Changes

Table 1: Temperature data for mixing equal masses of water

Liquid cold water plus hot water		
Cold water temperature before mixing (°C)	Hot water temperature before mixing (°C)	Mixture temperature (°C)
Solid water (ice) plus hot water		
Ice temperature before mixing (°C)	Hot water temperature before mixing (°C)	Mixture temperature (°C)
0		

2 Arguing from evidence

a. Given the actual hot and cold temperatures, what do you think the mixture temperature should have been if the ice could change to liquid (of the same temperature) without any change in energy?

b. Was the final temperature of the ice and water mixture about the same, more, or less than the final temperature of the water and water mixture?

c. Explain the difference in temperatures using the concepts of energy and phase change (heat of fusion). You may refer to the diagram on the right.

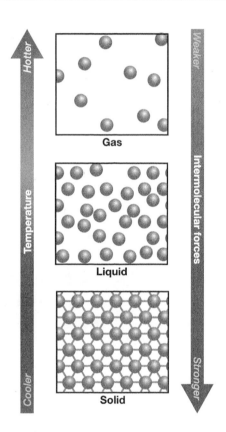

9.3 Exploring Heat Transfer

How do you analyze a system to see where heat transfer occurs?

There are many important examples of *heat transfer* in Earth's systems. Earth's internal energy is transferred through amazing convection cells in the mantle. Radiant heat from the Sun warms Earth. In this investigation, you will analyze a system for heat transfer through *conduction* and *convection* by making observations and collecting temperature data. Then, you will design your own system to observe heat transfer through thermal radiation.

Materials List
- Insulated beverage container (1-L capacity)
- Erlenmeyer flask (125 mL)
- 1- or 2-hole Rubber stopper to fit the flask
- Small amount of modeling clay
- Goggles and lab apron
- DataCollector
- Food coloring
- Stirring rod
- Temperature probe
- Hot and cold water
- Assorted materials for thermal radiation experiment
- 1-L Beverage pitcher

 Use caution when handling hot water.

1 Making predictions

You will set up a heat transfer system by placing a flask of hot water inside an insulated container of cold water. In the first part of the experiment, you will put a stopper with a hole in it on the hot water flask. In the second part of the experiment, you will do the same thing, but with the hole in the stopper closed off. In both parts of the experiment, you will monitor the temperature of the water in the insulated container.

a. Will the temperature of the cold water in the insulated container increase, decrease, or stay the same?

b. In the first part of the experiment, the hot water flask has a one- or two-hole stopper. In the second part, the hot water flask has a stopper with no holes. Will this make a difference in the temperature change of the cold water? Explain your reasoning.

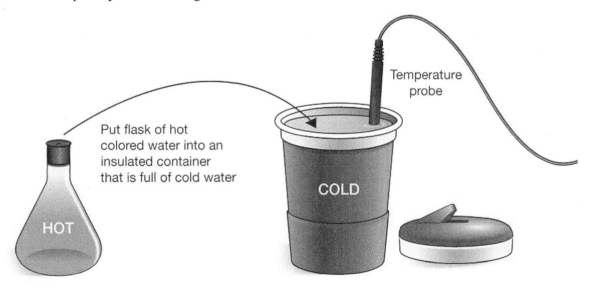

Investigation 9.3 Exploring Heat Transfer

2 Experiment part 1

When working with the flask of very hot water, be sure to use a hot mitt or tongs. You can easily use a regular thermometer for this investigation, but a temperature probe is even better (and more fun)!

1. Connect the temperature probe to the DataCollector and select Meter mode. Running time is displayed to keep track of elapsed time.
2. Remove the lid from the insulated beverage container. You won't need it for this experiment. Fill the container with 800 mL of cold water (12–14 °C is a good range). Record the temperature in Table 1.
3. Fill the 125-mL flask to the top with very hot water. Place several drops of food coloring in the water. Record the temperature of the hot water here so you can refer to it later: _____.
4. Quickly place the one- or two-hole stopper on the flask, and using a paper towel (to avoid hot water and stains from the dye), seal the stopper firmly in the neck of the flask.
5. Quickly place the flask in the insulated container. The water will be several inches above the flask. Tap on the stopper a few times to release any air bubbles.
6. Record the temperature of the water in the insulated container every 30 seconds for 300 seconds (5 minutes). Be sure to stir the water constantly for accurate temperature readings. Record the temperatures in Table 1.
7. Observe what is happening in the insulated container. You will sketch a diagram of your observations at the end of the experiment.

Safety Tip: Wear your goggles and use a hot mitt or tongs to hold the hot-water flask!

3 Observations: Experiment part 1

a. Soon after you placed the flask into the insulated container in part 1, you saw evidence of heat transfer taking place. Describe the evidence. Draw and label a sketch to illustrate your answer.
b. What type of heat transfer in part 1 of the experiment was the most visible: conduction, convection, or radiation? Explain.
c. What happened to the temperature of the water in the insulated container in part 1 of the experiment? How did this compare to your prediction?

4 Experiment part 2

1. Use some clay to seal up the holes in the rubber stopper.
2. Start with 800 mL of fresh cold water in the insulated container. Try to get the starting temperature as close to what it was in part 1 as you can.
3. Put fresh hot water in the flask, or reheat the water you used before. *The hot water temperature must be the same as it was in part 1*. Record the hot water temperature here: _____.
4. Repeat steps 4–6 from part 1 using the stopper that has its holes covered with clay.

5 Observations: Experiment part 2

a. Soon after you placed the flask into the insulated container in part 2, did you *see* evidence of heat transfer taking place? Explain.

b. What happened to the temperature of the water in the insulated container in part 2 of the experiment? How did this compare to your prediction?

c. Based on your knowledge of conduction, convection, and radiation, how was heat being transferred from the flask in part 2 to the water in the insulated container?

d. Draw a comprehensive sketch of how heat is being transferred in part 2 of the experiment.

Table 1: Temperature of water in insulated container

Time (s)	Part 1 temp (°C)	Part 2 temp (°C)
0		
30		
60		
90		
120		
150		
180		
210		
240		
270		
300		

6 Analyzing the temperature data

It is interesting to compare the change in temperature (called "delta T" and written ΔT) between parts 1 and 2 of the experiment.

a. What was the ΔT for part 1?

b. What was the ΔT for part 2?

c. Why was one ΔT significantly greater than the other? (Even a 2°C difference between the ΔT for parts 1 and 2 is significant when the overall ΔT is small.) Use your knowledge of conduction and convection to explain your answer.

Investigation 9.3 Exploring Heat Transfer

7. Designing a system to demonstrate heat transfer through thermal radiation

You have set up, performed, observed, and analyzed two systems to see where heat transfer occurs through convection and conduction. Your teacher will provide you with some materials. You and your group will need to come up with a system to demonstrate heat transfer through radiation. The goal is to develop a system that successfully demonstrates that heat can be transferred through the process of radiation.

To begin, focus on the main points of what you will be trying to show.

a. What is thermal radiation?

b. What happens during the process of thermal radiation?

c. What is the main difference between the other two forms of heat transfer, convection and conduction, and thermal radiation?

Use the answers to these three questions as a starting point for designing your demonstration. As you and your group go through the design process, allow everyone to voice any ideas they may have, list them all out, and then choose the ideas that have the best chance of being completed in the time frame you have been given by your teacher. Consider the following key points:

1. What has to happen for a successful demonstration of thermal radiation?
2. How can any measuring be done to prove that the demonstration has been successful?
3. How can you be sure that your design is safe for all those involved before performing the demonstration?
4. Is everyone in the group involved in each part of the design cycle, construction, performance, and analysis of your demonstration?
5. What contributions did you make to the group at each stage of the process from start to finish?
6. Can you perform a test run of your demonstration to be sure it works?
7. Are you sure any heat transfer that took place was due to thermal radiation and not convection or conduction somehow?
8. Did you analyze the heat transfer correctly?
9. What would be the best way to present your demonstration, the data you collected, and your findings in order to show that your system was successful in demonstrating thermal radiation?
10. Once you have completed the activity, answer this question: what would you have done differently to improve your system? What were your system's strong points?

With these key points in mind, collaborate with your group members and design a successful demonstration.

10.1 The Atom

How is an atom organized?

We once believed that atoms were the smallest units of matter. Then it was discovered that there are even smaller particles inside atoms! The structure of the atom is the underlying reason that nearly all the properties of matter we experience are what they are. This investigation will lead you through some challenging and fun games that illustrate how atoms are built from protons, neutrons, and electrons.

Materials List
- Atom Building Game

1 Modeling an atom

In the atom game, colored marbles represent the three kinds of particles. Red or green marbles are protons, blue marbles are neutrons, and yellow marbles are electrons.

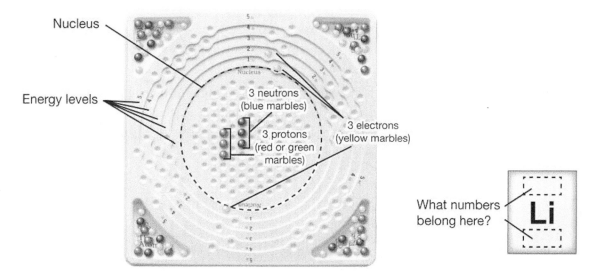

1. Build the atom above using three red or green, three blue, and three yellow marbles.
2. Fill in the blanks in the empty periodic table box for the atom you constructed.

2 Using your model

a. What is the number below the element symbol called? What does this number tell you about the atom?

b. What is the number above the element symbol called? What does this number tell you about the atom?

c. Why do some elements have more than one number above the symbol? What are the variations in this number called?

Investigation 10.1 The Atom

3 The Atomic Challenge

Atomic Challenge is a game that simulates the periodic table of elements.

The winner of the game is the first player to use all his or her marbles.

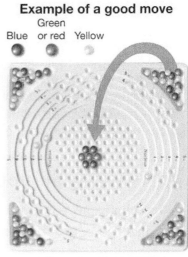

Example of a good move

$Li^7 + p + n + e = Be^9$

1. Each player should start with the following marbles: six blue marbles (neutrons), five red or green marbles (protons), and five yellow marbles (electrons).
2. Each player takes turns adding one to five marbles, but not more than five. The marbles may include any mixture of electrons, protons, and neutrons.
3. Marbles played in a turn are added to the marbles already in the atom.
4. If you add marbles that make an atom *not* shown on the periodic table, you have to take your marbles back and lose your turn. Only atoms in which the electrons, protons, and neutrons match one of the naturally-occurring elements on the table are allowed.
5. A player can trade marbles with the bank *instead* of taking a turn. The player can take as many marbles, and of as many colors, as needed, but must take at least as many total marbles as they put in. For example, a player can trade two yellows for one yellow, one blue, and one red or green.

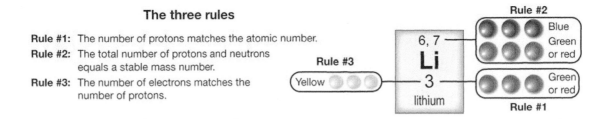

The three rules

Rule #1: The number of protons matches the atomic number.
Rule #2: The total number of protons and neutrons equals a stable mass number.
Rule #3: The number of electrons matches the number of protons.

4 Using the periodic table

Atoms that are not on the periodic table shown on the next page may exist in nature, but they are radioactive and unstable. For example, carbon-14 (C^{14}) is unstable and is not listed, although C^{12} and C^{13} are stable.

a. How many electrons does an atom of neon (Ne) have?
b. How many stable isotopes does oxygen (O) have?
c. Find one element on the periodic table that has no stable isotopes.
d. What element has atoms with 26 protons in the nucleus?
e. On most periodic tables, a single atomic mass is listed instead of the mass numbers for all the stable isotopes. How is this mass related to the different isotopes?

Periodic Table of the Elements 1-54
(Stable isotopes)

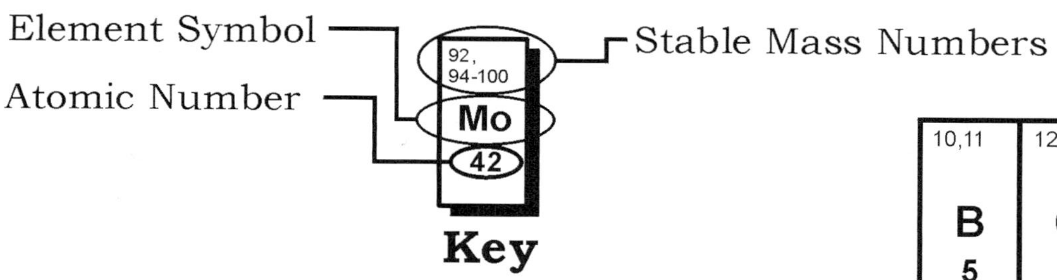

Key

| 42 | Mo | 92, 94-100 |

- Atomic Number
- Element Symbol
- Stable Mass Numbers

H 1 (1,2)																	He 2 (3,4)
Li 3 (6,7)	Be 4 (9)											B 5 (10,11)	C 6 (12,13)	N 7 (14,15)	O 8 (16-18)	F 9 (19)	Ne 10 (20-22)
Na 11 (23)	Mg 12 (24-26)											Al 13 (27)	Si 14 (28-30)	P 15 (31)	S 16 (32-34, 36)	Cl 17 (35,37)	Ar 18 (36,38, 40)
K 19 (39,41)	Ca 20 (40,42, 44,46, 48)	Sc 21 (45)	Ti 22 (46-50)	V 23 (51)	Cr 24 (50, 52-54)	Mn 25 (55)	Fe 26 (54,56-58)	Co 27 (59)	Ni 28 (58,60-62,64)	Cu 29 (63,65)	Zn 30 (64,66-68,70)	Ga 31 (69,71)	Ge 32 (70,72-74,76)	As 33 (75)	Se 34 (74,76-78,80,82)	Br 35 (79,81)	Kr 36 (78,80, 82-84, 86)
Rb 37 (85)	Sr 38 (84, 86-88)	Y 39 (89)	Zr 40 (90-92, 94,96)	Nb 41 (93)	Mo 42 (92, 94-100)	Tc 43 (none)	Ru 44 (96,98-103,104)	Rh 45 (103)	Pd 46 (102,104-106,108,110)	Ag 47 (107,109)	Cd 48 (106,108, 110-112, 114,116)	In 49 (113)	Sn 50 (112,114-120,122, 124)	Sb 51 (121)	Te 52 (120,122, 124-126, 128,130)	I 53 (127)	Xe 54 (124,126, 128-132, 134,136)

10.2 Energy and the Quantum Theory

How do atoms absorb and emit light energy?

The electrons in an atom are organized into energy levels. You can think of energy levels like a staircase where the electrons can be on one step or another, but cannot exist in between steps. When an electron changes levels, the atom absorbs or emits energy, often in the form of light. This investigation will teach you a challenging and fun game that simulates how atoms exchange energy through light: the process by which a laser works.

Materials List
- Atom Building Game
- Spectrometer

1 The neon atom

1. Build a neon atom with 10 each of protons (red or green marbles), neutrons (blue marbles), and electrons (yellow marbles).
2. Set the electrons in the lowest spaces possible.
3. Find the following cards in the Photons and Lasers can
    - Pump 1 (red)
    - Pump 2 (yellow)
    - Laser 1 (red)

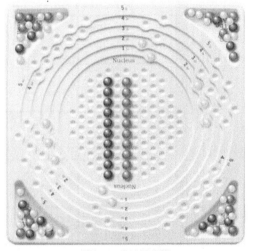

Neon-20 (Ne^{20})
10 protons, 10 neutrons, 10 electrons

2 How atoms exchange energy

a. Explain the meaning of the term *ground state* when applied to an atom.

b. Can the second energy level of neon hold any more electrons? How does this affect neon's chemical properties and position on the periodic table?

c. Take the red "pump 1" card from your hand and put it on the atom board. Move one electron from level 2 to level 3. Explain what this sequence of actions represents in a real atom.

d. Take the yellow "pump 2" card from your hand and put it on the atom board. Move any one electron up two levels. Explain what this sequence of actions represents in a real atom.

e. Take the red "laser 1" card from your hand and put it on the atom board. Move any one electron down one level. Explain what this sequence of actions represents in a real atom.

3 The photons and lasers game

Photons and Lasers Card Deck

Pump Cards Add energy to the atom and advance electrons up levels

Laser Cards Release energy from the atom and drop electrons down levels

1. The first player to reach 10 points wins the game.
2. Each player starts with five cards and plays one per turn. A new card is drawn each turn to maintain a hand of five.
3. Playing a pump card allows the player to advance one electron level up by the number of levels shown on the card (1–4). No points are scored by playing pump cards.
4. Playing a laser card allows the player to drop electrons one level lower. The player scores one point per electron per level. For example, moving two electrons down two levels scores four points.
5. Rules for playing laser cards:
    Electrons can only be moved down if there are empty states for them to move to.
    Electrons can only be moved from one level in a turn.
    If the card says "laser 2" then each electron must move 2 levels.

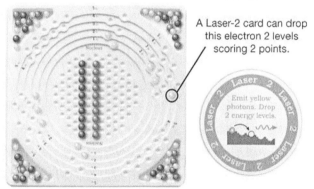

A Laser-2 card can drop this electron 2 levels scoring 2 points.

4 Constructing explanations

a. What does the term *excited state* mean with respect to energy and atoms?

b. What physical principle prevents two electrons from moving into the same state?

c. In order of increasing energy, arrange the following colors of light: blue, red, green, and yellow.

d. Could an atom emit one photon of blue light after absorbing only one photon of red light? Explain why or why not.

e. Suppose a real atom had energy levels just like the game. Could this atom make blue-green light with an energy in between blue and green? Explain what colors this atom could make.
Take the spectrometer and look at the light from a fluorescent lamp. It looks white but you will see lines of certain colors. The lines are proof that electrons in atoms really do have energy levels.

10.3 Nuclear Reactions and Radioactivity

How do nuclear changes involve energy?

You would be very surprised to see a bus spontaneously transform into three cars and a motorcycle. But radioactive atoms do something very similar. If left alone, a radioactive atom eventually turns into another kind of atom with completely different properties. This investigation looks at some basic concepts behind radioactivity.

Materials List
- Atom Building Game
- 50 Pennies
- Cup
- Graph paper

1 Radioactivity

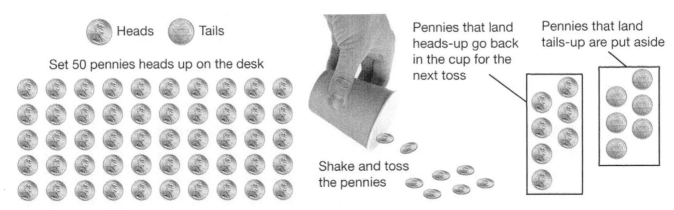

1. Place 50 pennies in a paper cup, shake them, and dump them onto the table. Each penny represents an atom of carbon-14.
2. Separate the pennies that land tails-up. Count the heads-up pennies and tails-up pennies and record the number of each in Table 1 in the row for the first toss.
3. *Put only the pennies that landed heads-up back in the cup.* Put the tails-up pennies aside. Shake the cup again and dump the pennies on the table.
4. Record the number of heads-up and tails-up pennies in the row for the second toss.
5. Repeat the experiment using only the pennies that landed heads-up until you have one or no pennies left.

Table 1: Coin toss decay simulation

	Heads-up pennies	Tails-up pennies
Start	50	0
First toss		
Second toss		
Third toss		
Fourth toss		
Fifth toss		
Sixth toss		
Seventh toss		
Eighth toss		

Nuclear Reactions and Radioactivity — Investigation 10.3

2. Constructing explanations

a. Make a graph showing the number of heads-up pennies on the *y*-axis and the number of tosses on the *x*-axis (0, 1, 2, 3,...).

b. On average, what percentage of pennies are lost on each toss? *Lost* means they came up tails and were removed.

c. How does the concept of half-life relate to the experiment with pennies? What does one half-life correspond to?

3. Build a radioactive atom

1. Build a carbon-14 atom (C^{14}). This isotope of carbon is radioactive.
2. Take one neutron out and replace it with a proton and an electron. This is what happens in radioactive decay of C^{14}.

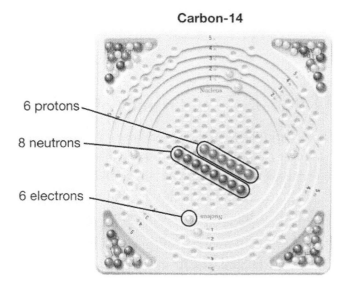

Carbon-14

6 protons
8 neutrons
6 electrons

4. Arguing from evidence

a. Research what happens to C^{14} when it decays. What element does it become? What particles are given off?

b. What is the average time it takes for 50 percent of the C^{14} atoms in a sample to decay?

c. Suppose you have 50 atoms of C^{14} and you watch them for a very long time. How do the results of your penny-flipping experiment describe the number of C^{14} atoms?

d. We actually find C^{14} in the environment. Research where it comes from.

e. Describe two other types of radioactivity and give an example of each.

f. *Challenge:* You cannot predict when any one atom will decay, just as you cannot predict whether a penny will come up heads or tails. Why can you predict that 50 percent of the C^{14} atoms will decay every half-life?

11.1 Frames of Reference

How does your frame of reference affect what you observe?

The word *relativity* refers to the idea that what you experience depends on the relative motion of your frame of reference. This investigation about frames of reference will give you a clue to the meaning of relativity.

In this investigation, you will:

- explore the concept of "frame of reference."
- create and observe two frames of reference for the same event.

Materials List
- SmartTrack
- Energy Car
- Photogate
- DataCollector
- Rolling table
- Masking tape
- Meter stick or tape measure

1 Relativity and frames of reference

The word *relative* in the special theory of relativity means that velocities can only be determined relative to a *reference frame*. It makes no sense in physics to say an object is moving at 1 m/s without providing a reference. For example, you might measure a speed of 1 m/s relative to your lab or classroom. However, Earth rotates and also circles the Sun. A velocity of 1 m/s relative to the lab (on Earth) is *not* the same as a velocity of 1 m/s relative to the Sun.

Consider the following experiment. A person on a train throws a dart at a dart board. Relative to the train, the dart travels a distance of 5 meters to the dart board in 0.5 seconds. The train is moving forward at a speed of 30 m/s.

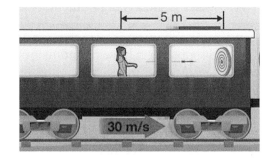

a. There are two reference frames important for understanding the motion of the dart. What are they?

b. What is the speed of the dart relative to the train?

c. What is the speed of the dart relative to the ground?

Frames of Reference   Investigation 11.1

2 Demonstrating two frames of reference

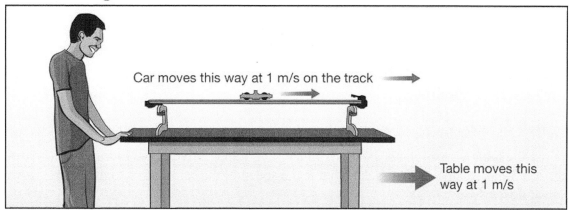

1. Set the SmartTrack up and make it level on the rolling table.
2. Attach two launchers on opposite ends of the track facing each other. Put a rubber band onto the arms of both of the launchers in an x pattern by giving the rubber band a half-twist, then secure it in place between the retaining washers with the securing screws.
3. Plug the photogate into the A slot on the DataCollector and place the photogate in the middle of the track. Use the DataCollector in Timer mode with the Interval function to measure the speed of the car when it is launched by the rubber band.
4. Adjust the launcher to propel the car at an approximately constant speed of 1 m/s. Once it is set, remove the photogate and DataCollector from the area around the rolling table.
5. Mark a distance of 3–4 meters on the floor with tape. The table will be observed while being rolled across this area.
6. Use the stopwatch mode of the DataCollector to calibrate your pace so you can push the rolling table at a constant speed of 1 m/s. Don't worry about getting things exact; this is only a demonstration of an idea.
7. The person pushing the rolling table will launch the car so it races down the track and bounces off the rubber band at the far end to return the way it has come. This person should look only at the car on the track. The rest of the group should stand to the side and watch the car from their perspective.
8. Everyone should describe the motion of the car based on their frame of reference before and after the car bounces off the rubber band. The car may bounce back and forth a few times. Record everyone's observations.

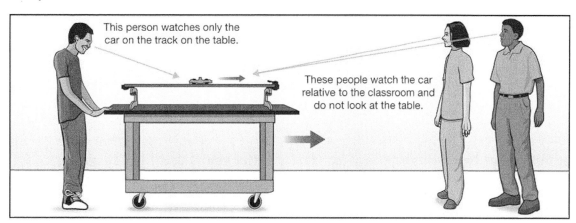

Investigation 11.1 Frames of Reference

3 Constructing explanations

a. Imagine that the person pushing the table was in a box, able only to see the car and track and nothing outside the box. What motion of the car do they see relative to themselves?

b. What motion of the car do the outside observers see relative to themselves?

c. There are two important frames of reference in this demonstration. What are they?

d. One frame of reference is in motion relative to the other. What is this motion?

e. Suppose the table was rolling at exactly 1 m/s to the right relative to the room, and the car was rolling to the left 3 m/s relative to the track. What is the speed of the car relative to the room? How did you arrive at this answer?

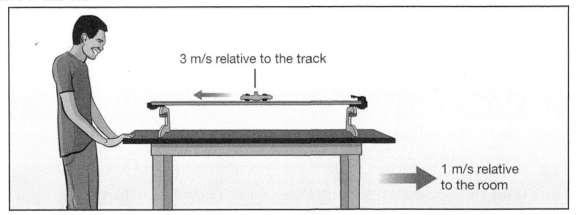

4 A thought experiment

Imagine you are watching a train from a distance. You see two bolts of lightning hit the train at the same time. To you, the two events (lightning strikes) are *simultaneous* because it takes the same amount of time for light from either event to reach you.

If you were sitting on the train however, the situation looks different. Suppose you are in the center of the train. If the train were at rest, you would see two simultaneous lightning strikes. But, since the train is moving, it moves between the time the lightning hits and when light from the lightning reaches you.

a. When you are on the moving train, do you observe the lightning hit the front of the train first, the back first, or does light from both lightning strikes reach you at the exact same time?

b. (Discussion question) Explain the reasoning behind your answer to question a above.

Whether two events occur at the same time depends on the relative motion of your frame of reference.

According to Einstein, two events happen at the same time (for you) if the light from each reaches you at the same time. This is an interesting meaning for "at the same time." It suggests that *time itself* is dependent on your frame of reference. Many people find this idea very strange, but it is true. Time passes at different rates in reference frames that are moving relative to each other.

11.2 Special Relativity

What are some of the implications of special relativity?

Time moves more slowly for an object in motion than it does for an object that is not in motion. In practical terms, clocks run slower on moving spaceships compared to clocks on the ground. If a spaceship is moving very fast, it is possible for one year to pass on the spaceship while 100 years pass on the ground. This effect is known as *time dilation*.

In this investigation, you will

- explore some consequences of time dilation.
- calculate the equivalence of mass and energy using Einstein's formula, $E = mc^2$.

Materials List
- Access to the internet and other research materials
- Simple calculator

1 An imaginary experiment with light

With ordinary speeds, the speed *you* measure depends on the relative motion of *your* frame of reference. The fundamental rule of the theory of relativity is that the speed of light is always the same *even if your frame of reference is moving*. This single idea caused Einstein to change all our concepts of space and time.

Einstein imagined having a clock that measures time by counting the trips of a beam of light going up and down between two mirrors. The clock is on a moving spaceship. A person standing next to the clock sees the light go straight up and down. The time it takes to make one trip is the distance between the mirrors divided by the speed of light.

Special relativity says the speed of light is the same for every observer, regardless of relative motion.

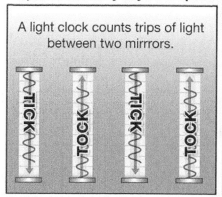

A light clock counts trips of light between two mirrrors.

In the ship the light goes straight up and down.

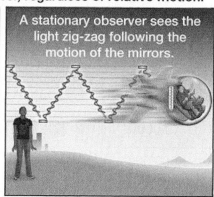

A stationary observer sees the light zig-zag following the motion of the mirrors.

To someone who is not moving, the path of the light is not straight up and down. The light appears to make a zigzag because the mirrors move with the spaceship.

This would not be a problem, except that the speed of light must be the same to all observers, regardless of their motion. How can this be?

Investigation 11.2 Special Relativity

2 Constructing explanations

The following are all questions for class discussion. They are challenging!

a. Describe the two reference frames that are important. Who is in each one?

b. What is the relative motion between the two reference frames?

c. Does the person on the ground see the light travel a distance that is longer, shorter, or the same compared to the distance seen by the person watching the light on the spaceship?

d. Prior to Einstein, speed was always calculated as the distance traveled divided by the time taken. Thinking this way, does the person on the ground see the light in the clock move faster, slower, or at the same speed compared to what the person in the spaceship sees?

e. The theory of relativity requires that the speed of light be the same for all observers, regardless of relative motion. If the speed is the same, but the distance is different, what other variable must also be different in the two reference frames?

f. Do clocks on board the spaceship run slower, faster, or at the same rate compared to clocks on the ground?

3 When does special relativity become important?

Table 1 gives some examples of how time dilation affects the perception of one second of time for observers moving relative to each other at different speeds.

Table 1: Speed and Time Dilation

Speed			Time for the moving observer	Time for the observer at rest
m/s	mph	% of c	(s)	(s)
100	224	3.3×10^{-5}	1.0000	1.0000
1,000	2,240	3.3×10^{-4}	1.0000	1.0000
10,000	22,400	0.0033	1.0000	1.0000
1,000,000	2.24×10^6	0.33	1.0000	1.0000
1.00×10^7	2.24×10^8	3.3	1.0000	1.0006
1.00×10^8	2.24×10^8	33	1.0000	1.0607
2.00×10^8	2.24×10^8	67	1.0000	1.3424
2.80×10^8	6.26×10^8	93	1.0000	2.7981
2.90×10^8	6.49×10^8	97	1.0000	3.9434
2.99×10^8	6.69×10^8	99.7	1.0000	13.6976

a. A high-performance aircraft flies at a speed of 1,340 m/s, or four times faster than the speed of sound (340 m/s). At this high speed, can the effects of time dilation be perceived by a person with an ordinary watch?

b. A rocket traveling to Mars must have a speed greater than the minimum speed required to break Earth's gravitational attraction. This minimum speed is called the escape velocity. Research to find Earth's escape velocity. Is the escape velocity fast enough that relativity must be considered for normal purposes such as synchronizing two clocks?

c. The numbers in Table 1 were calculated using a formula proposed by Einstein. Research the formula to identify what the variables mean.

4 The equivalence of mass and energy

EINSTEIN'S MASS-ENERGY FORMULA

$$E = mc^2$$

Energy (J) — E; Mass (kg) — m; Speed of light (3×10^8 m/s) — c

According to Einstein's theory of special relativity, mass and energy are two forms of the same thing. One way to think about mass is as extremely concentrated energy. In fact, according to the mass-energy formula (above), 1 kilogram of mass is equivalent to 9×10^{16} joules of energy.

a. Calculate the amount of energy used by a 100 W light bulb that is left on continuously for one year.

b. Suppose you could extract all of the energy in one kilogram of mass with 100 percent efficiency. How long could this amount of energy keep the 100 W light bulb lit? Give your answer in years.

c. Nuclear reactors convert a tiny fraction of the mass of uranium atoms into energy. Assume that 0.07 percent of the mass of uranium is converted to energy. How much energy do you get from one kilogram of uranium in a nuclear reactor?

d. To appreciate the energy obtained from one kilogram of uranium, estimate the electric power used by a city of one million people. Assume that each person in the city uses an amount of electricity equal to 10 light bulbs that use 100 watts each. Calculate the total energy used by multiplying the total power in watts by the number of seconds in one year (you will get a very large number).

e. How many kilograms of uranium must be used in a nuclear reactor to produce this amount of energy?

f. One gallon of ordinary gasoline yields about 1.3×10^7 joules of energy. Calculate how many gallons of gasoline are required to produce the amount of energy you estimated in part d.

11.3 General Relativity

What is Einstein's theory of general relativity?

Consider rolling a ball across a sheet of graph paper. If the graph paper is flat, the ball rolls along a straight line. A flat sheet of graph paper is like "flat space." In flat space, parallel lines never meet, and all three angles of a triangle add up to 180 degrees. Flat space is what you would consider "normal." In this investigation, you will examine how Einstein's theory of general relativity challenges the idea of "normal" space.

Materials List
- Access to the internet and other research materials

1 General relativity

A large mass, like a star, curves space near itself. The middle diagram to the right shows an example of a graph made of rubber that holds a large mass. The large mass has created a "well" on the graph; notice the curved lines. If you roll a ball along this graph, its path bends as it rolls near this place where the graph is stretched by the mass. From directly overhead, the graph grid still looks like squares. If you look straight down on the graph, the path of the ball appears to be deflected by a force pulling it toward the large mass. You might say the ball "felt" a force of gravity which deflected its motion. You would be right. This effect of curved space is identical to the force of gravity.

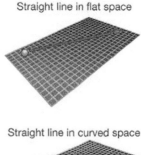
Straight line in flat space

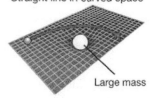
Straight line in curved space
Large mass

Observed force of gravity

a. Suppose "straight line" is defined as "the path of motion an object follows when the net force acting on it is zero." Does this definition describe the same thing as your idea of a "straight line"?

b. Suppose you stretch a rubber sheet and place a heavy glass ball in the center. The glass ball depresses the sheet a few centimeters. A plastic ball is rolled in a straight line from the edge of the sheet as shown in the diagram. Sketch the path of the plastic ball as it passes by the glass ball.

c. Suppose the rubber sheet was completely transparent, and you could not see the glass ball either. Propose an explanation that would explain the observed path of the plastic ball using the idea of force.

2 Einstein's theory

Einstein's general theory of relativity describes how objects with mass distort the very shape of space itself.

a. Do some research and look for the connection between gravity and Einstein's theory of general relativity.

b. Come up with your own description of space-time.

c. Research black holes. What is their connection with Einstein's general theory of relativity?

d. Can you come up with a classroom activity that would demonstrate Einstein's general theory of relativity?

12.1 Measuring Voltage and Current

How do you measure voltage and current in electric circuits?

We use electricity nearly every minute of every day. In this investigation you will build simple circuits and learn how to measure current and voltage, two fundamental quantities that describe electricity.

Materials List

- Electric Circuits kit
- Two "D" batteries
- Multimeter
- Metal paper clip, plastic straw, piece of string, rubber band, pen cap, or other similar objects

1 Lighting a bulb

In the first part of the investigation, you are going to be building simple circuits *without* the use of the battery and bulb holders. While figuring out how to make your circuits, it is important that you do not create a short circuit. Connecting the two ends of the battery to each other without the light bulb in between creates a short circuit that will quickly drain the battery.

a. Closely examine the bulb (without its holder). Draw a diagram that shows all of its different parts. Your diagram should be drawn so it is larger than the actual bulb.

b. Use two wires, one bulb, and one battery (no holders) to make a circuit that lights the bulb. Make a drawing that shows how you connected the different components of the circuit. Clearly show which parts of the bulb and battery were connected.

c. Repeat the previous step, but use only one wire this time.

d. Draw three ways to connect the battery, bulb, and wire(s) that will *not* make the bulb light.

2 Building a circuit

1. Build the circuit shown in the photo with one battery, a switch, and a bulb.
2. Open and close the switch and see what happens.

3 Constructing explanations

a. How can you tell electric current is flowing in the circuit? Can you see the current?

b. Current flows from positive to negative. Trace the flow of current around the circuit with your finger. Unscrew the bulb and look in the holder to see how it connects to the bulb. Describe the path the current follows.

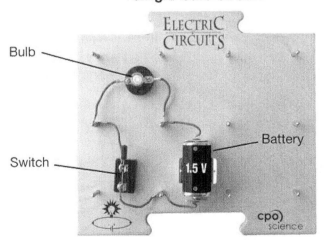

Single bulb circuit

c. How does the switch cause the current to stop flowing?

d. Why does the bulb go out when you open the switch?

4 Conductors and insulators

Materials through which electric current flows easily are called *conductors*.

Materials through which current does not flow easily are called *insulators*.

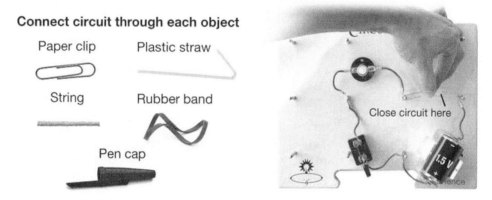

1. Break one connection in your one-bulb circuit.
2. Complete the circuit by touching different materials between the wire and the post.
3. Notice which materials allow the bulb to light and which do not.

5 Constructing explanations

a. Make a table listing the materials as either conductors or insulators.

b. What characteristics are shared by the conductors you found?

c. What characteristics are shared by the insulators you found?

6 Circuit diagrams

For describing electric circuits, we use the language of *circuit diagrams*. In a circuit diagram, wires are represented by solid lines. Electrical devices like switches, batteries, and bulbs are represented by symbols.

a. Using these symbols, draw a picture of the circuit you built with one battery, switch, and light bulb.

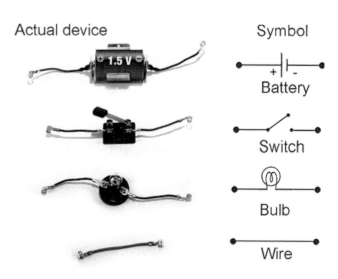

7 Measuring the voltage of a battery

Set your multimeter to measure DC volts. The red lead connects to the positive terminal and the black lead connects to the negative terminal.

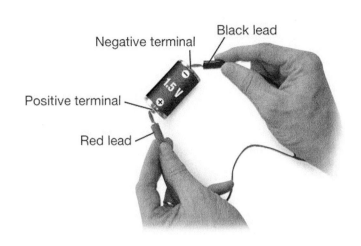

a. Measure the voltage of one battery and record your reading. Repeat for a second battery.

b. Connect the two batteries so the positive end of one is touching the negative end of the other. Measure the voltage of the two-battery combination. Record your measurement.

c. Flip the batteries over so the other sides are connected but are still positive to negative. Record the total voltage.

d. Position the batteries so the two positive ends are touching. Record the total voltage of the two-battery combination. Both the red and black leads will be touching negative terminals, so you may get a negative voltage reading. Repeat with the two negative ends touching. Record the voltage.

e. Explain what your measurements show about the relationship between the individual voltages and the total voltage when the batteries are connected in different ways.

8 Measuring current

You must connect the meter differently to measure current. Instead of reading a difference between two points as for voltage, the meter measures the current passing through it. You must force the current to flow through the meter by eliminating all other paths the current could take. Carefully follow the instructions below so you do not damage the meter.

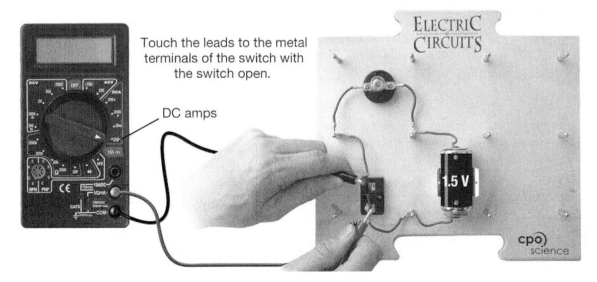

1. Set the multimeter to measure DC amps (current).
2. Open the switch. The bulb should go out. Touch the red lead of the meter to the metal part of the switch closest to the battery's positive terminal (+).
3. Touch the black lead of the meter to the metal part on the other side of the switch.

Investigation 12.1 Measuring Voltage and Current

4. The bulb should light, showing you that current is flowing through the meter. The meter should display the current in amps. This is the current flowing around the circuit carrying energy from the battery to the bulb. Remove the meter.

a. How much current is flowing in the circuit when the bulb is making light?

9 A circuit with a dimmer switch

The potentiometer (or *pot*) is an electrical device that can be used to make a dimmer switch. When the dial on the potentiometer is turned one way, it acts like a closed switch and current flows freely through it. When the dial is turned the other way, it resists the flow of current.

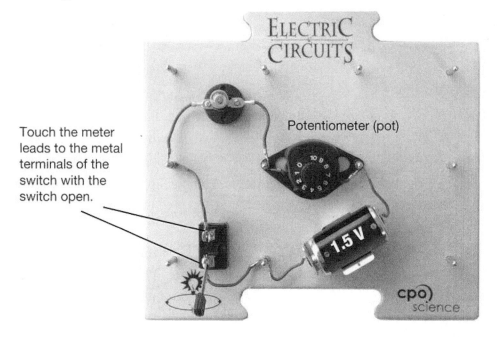

1. Connect the circuit as shown in the diagram using the potentiometer, a battery, wire, and a bulb.
2. Adjust the dial and watch what happens to the bulb.
3. Use the meter to measure the current in the circuit for four different settings of the potentiometer. Record your data in Table 1.

Table 1: Potentiometer settings and current

Pot dial position	Current in circuit (A)	Observed light output of bulb

10 Constructing explanations

a. As you increased the setting of the potentiometer, what happened to the current in the circuit?
b. Explain the relationship between the current and the brightness of the bulb.
c. Describe a use for a potentiometer other than as a dimmer switch for a light.

12.2 Resistance and Ohm's Law

What is the relationship between current and voltage in a circuit?

Electrical devices get the energy they use from the current that flows through them. When designing an electrical device or a circuit, it is important for the proper amount of current to flow for the voltage that is available. Resistance is the property of electricity that helps regulate the current in the circuit. You will explore resistance and Ohm's law, the equation that relates voltage, current, and resistance.

Materials List
- Electric Circuits kit
- 2 "D" batteries
- Multimeter

1 Mystery resistors

A *resistor* is used in a circuit to provide resistance. You have green, blue, and red resistors in your kit with values of 5 Ω, 10 Ω, and 20 Ω, but you don't know which is which.

1. Make a circuit with a switch, battery, and resistor, as pictured above.
2. Set the meter to measure current (DC amps). Open the switch. Measure the current by connecting the meter across the open terminals of the switch.
3. Set the meter to measure voltage (DC volts). Close the switch. Measure the voltage across the mystery resistor as pictured above.
4. Repeat the current and voltage measurements for each of the mystery resistors.

Table 1: Resistor Currents

Resistor color	Voltage across resistor (V)	Current (A)
Green		
Blue		
Red		

a. Use your knowledge of Ohm's law to determine the value of each resistor. The resistance you calculate from Ohm's law will not come out exactly to 5, 10, or 20, because the meter itself has a small resistance.

Investigation 12.2 Resistance and Ohm's Law

2 Resistance and potentiometers (pots)

The potentiometer (pot) you used in the previous investigation is really a *variable resistor*. A variable resistor allows you to change its resistance by turning a dial.

1. Set the multimeter to measure resistance (Ω). Use the meter to measure the resistance of the pot for different positions of the dial. The pot should not be in the circuit; you can just touch the multimeter leads across the pot.
2. Take your first reading with the pot turned all the way to the left. Then take four or five readings until the pot is turned all the way to the right.

Table 2: Pot settings and resistance

Pot dial position	Resistance (Ω)

3 Voltage in a dimmer circuit

When you measure voltage in a circuit, you are measuring the change in the energy carried by the current at different points in the circuit. The energy carried by the current is reduced whenever current flows through a device that has a resistance greater than zero.

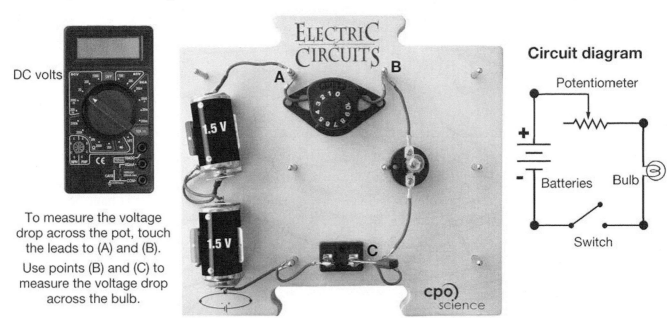

DC volts

To measure the voltage drop across the pot, touch the leads to (A) and (B).

Use points (B) and (C) to measure the voltage drop across the bulb.

Circuit diagram

1. Build the dimmer circuit with two batteries, the pot, a switch, and a bulb. Make sure the batteries are connected so the positive end of one is facing the negative end of the other.

2. Measure the voltage across the pot and across the bulb for five different dial settings of the pot. Record the observed light output at each dial position measured.

Table 3: Pot settings and voltage drops

Pot dial position	Voltage across pot (V)	Voltage across bulb (V)	Observed light output

4 Arguing from evidence

a. What happens to the voltage across the bulb as the voltage across the pot increases?

b. Calculate the sum of the bulb voltage and pot voltage for each of the five dial positions.

c. What do you notice about the sum of the voltages?

d. How do you think the voltages would compare if one battery were used instead of two?

e. Suppose you measured the current in the circuit for each of the five dial settings. Which dial setting would cause the current in the circuit to be the least? Why? Use Ohm's law in your explanation.

12.3 Building an Electric Circuit Game

How can you use electricity to test your manual dexterity?

Do you have a steady hand? In this project, you will create the setup for a game that challenges your manual dexterity. Can you move a small loop of wire over a complicated maze without tripping the light bulb? Try it and see!

Materials List
- Electric Circuits kit
- 1 "D" battery
- 1 m 12-gauge copper wire (not insulated)
- 50-cm Piece of 16-gauge insulated copper wire
- Electrical tape
- Wire stripper tool
- Permanent-ink marking pen
- Metric ruler or measuring tape

1 Building the game

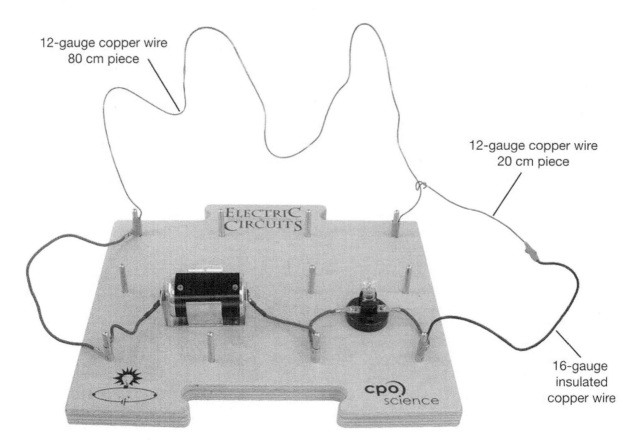

1. Place the battery, light bulb, and long connecting wire on the circuit board as shown above.
2. Cut a 20-centimeter piece from one end of the 1-meter piece of 12-gauge copper wire.

3. Bend one end of the 20-centimeter piece into a loop with a diameter no larger than a dime. The smaller the loop, the more challenging the game! Twist the wire to secure the loop. You have just constructed the wand for your game board.

4. Strip two centimetes of plastic coating from each end of the 50-centimeter length of 16-gauge wire. (Your teacher may help with this part.)

5. Wrap one end of the exposed wire around the base of your wand and secure with electrical tape.

6. Wrap the other end of the exposed-copper wire around the right front corner post of the electricity table. (The light bulb wire should also be connected to this post.) Secure with electrical tape.

7. Measure 15 centimeters in from each end of your remaining 80-centimeter piece of 12-gauge copper wire. Mark the two spots with permanent ink. *Do not* cut the wire.

8. Make a 90° bend in the wire at each spot so that the wire is shaped like a wide, upside-down *U*.

9. Bend the long horizontal section of the wire into a series of hills and valleys (see illustration). Adjust the bends until the two 15 cm "legs" of the wire are 23 cm apart.

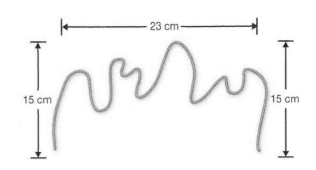

10. Place one of the 15-cm "legs" alongside the left, rear post of the electricity grid. The long connecting wire should be attached to this post. Secure the leg with electrical tape.

11. Slide the loop of your wand over the other leg of the 12-gauge wire.

12. Use electrical tape to secure this leg to the right, rear post of the electricity grid. Make sure that the tape covers the entire post.

13. Make sure that the loop in the wand will slide down the post. The loop should be placed in this position when the game is not in use.

2 Playing the game

Now you are ready to play! Using one hand, move the loop in the wand over the hills and valleys—but don't let the loop touch the copper wire! Try to make it all the way across without lighting the bulb.

Variation: Inexpensive buzzers can be purchased at electronic or hobby stores and placed in the circuit alongside the bulb.

3 Constructing explanations

a. Why was copper wire used to make the wand and the part with the hills and valleys?

b. Why were you instructed to strip the coating from the insulated wire?

c. How could you set up the circuit in the game so the light would be lit at the beginning and then would go out if you touched the wand to the copper wire? Draw a diagram to show your solution and explain why it works.

13.1 Series Circuits

How can devices be connected in circuits?

A simple electric circuit contains one electrical device, a battery, and a switch. Flashlights use this type of circuit. However, most electrical systems, such as televisions and computers, contain many electrical devices connected together in multiple circuits. This investigation introduces one way to connect multiple devices in a circuit.

Materials List
- Electric Circuits kit
- Multimeter
- 2 "D" batteries

1 Series circuits

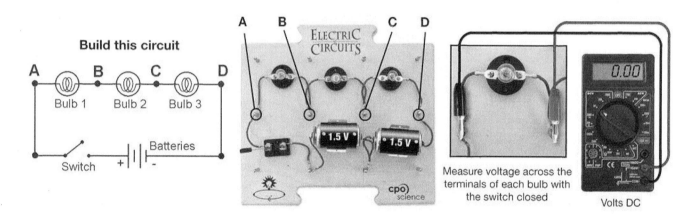

1. Using two batteries, build the simple circuit with three light bulbs and a switch as shown in the circuit diagram above.
2. Set the meter to DC volts. Close the switch and measure the voltage across the different places by touching the meter's leads to the bulbs' terminals. Record the voltages in Table 1.

Table 1: Voltage measurements (volts)

Between A and B (V)	Between B and C (V)	Between C and D (V)	Between A and D (V)

2 Analyzing your data

a. What relationships do you see among the voltage measurements in Table 1?

b. The batteries are supposed to be 1.5 volts each. Does this agree with your data?

c. What do the voltage measurements tell you about the energy in the circuit?

Series Circuits   Investigation 13.1

3 Current in series circuits

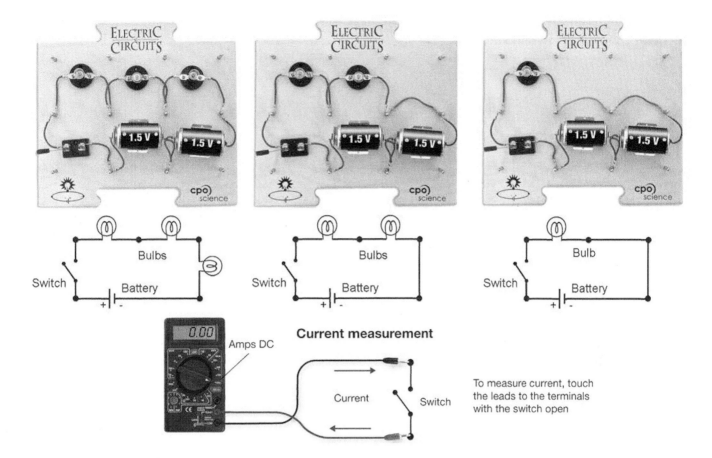

1. Set the meter to DC amps. Measure the current by opening the switch and touching the leads of the meter to the terminals of the switch in the three-bulb circuit. Record your measurements in Table 2.
2. Remove one bulb and replace it with a wire. Measure and record the current for the two-bulb circuit.
3. Remove a second bulb and replace it with a wire. Measure and record the current again for the one-bulb circuit.

Table 2: Current measurements (amps)

Three bulbs (A)	Two bulbs (A)	One bulb (A)

4 Constructing explanations

a. What happens to the current in the circuit as the number of bulbs is reduced? Explain why this occurs using Ohm's law and the concept of resistance.

b. What happens to the other two bulbs when one bulb is unscrewed from the three-bulb circuit? Try it and explain why the circuit behaves as it does.

Investigation 13.1 Series Circuits

5 Short circuits

A short circuit is an easy (but dangerous) shortcut that current can travel through to avoid one or more of the electrical components in the circuit.

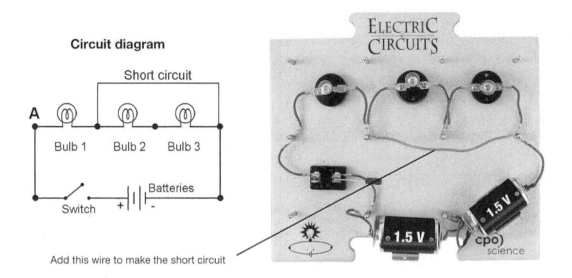

1. Rebuild your three-bulb circuit with the switch open.
2. Measure the current and observe how bright the bulbs are. Record the current in Table 3.
3. Add a section of wire that bridges the last two bulbs in the circuit as shown in the right section of the picture above. This wire is the "short circuit."
4. Complete the circuit (with the switch open) using the meter to measure the current. Observe which bulbs light and how bright they are. Record the current in Table 3.

Table 3: Short circuit current measurements (amps)

Three bulbs in series (A)	Three bulbs with short circuit (A)

6 Constructing explanations

a. Compare the current in the three-bulb circuit with the current when two bulbs are bypassed by a short circuit. Which is greater? Use Ohm's law and the concept of resistance to explain why.

b. How does the current in the "short circuit" version compare with the current you measured in a one-bulb circuit? Explain why this should be true.

c. How does the resistance of a wire compare to the resistance of a bulb? You may want to measure the resistances to test your answer. (*Note*: Most meters cannot measure very low resistance and display "0.00" when the resistance is lower than 0.01 Ω.)

d. Why would a short circuit be dangerous? Discuss the consequences of very large currents in wires.

13.2 Parallel Circuits

How do parallel circuits work?

In the last investigation, you learned how to measure current and voltage in series circuits. In this investigation, you will build parallel circuits and measure their electrical quantities. Parallel circuits are complex and are not intuitive, so you will need to examine your data carefully to understand how these circuits work.

Materials List
- Electric Circuits kit
- 2 "D" batteries (fresh)
- Multimeter

1 Building circuits

You should recall from the previous investigation that bulbs in series get dimmer as more are connected to the same batteries. In this investigation, you are going to figure out how to build circuits to brightly light more than one bulb at a time.

a. Build a simple circuit on the circuit board using two batteries, a light bulb, and some wires. The batteries and bulb should be in their holders. Use symbols to draw a circuit diagram for your circuit.

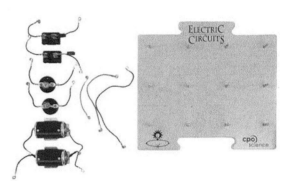

b. Notice how bright the bulb is. Now add a second bulb to your circuit so that both bulbs are approximately as bright as the one bulb in the previous step. Ask your teacher to check your circuit to make sure you built it correctly. Draw a circuit diagram for your circuit.

c. Add two switches to your circuit. Opening one of the switches should turn both bulbs off. Opening the other switch should turn only one bulb off. This is tricky! Ask your teacher to check that your circuit is correct, and then draw a circuit diagram.

2 Constructing explanations

a. The two-bulb circuit you built is called a *parallel circuit*. Explain how a parallel circuit is different from a series circuit.

b. Close the switches so both of the bulbs are on. Unscrew one of the bulbs from its holder. Describe what happens to the other bulb.

c. Do you think all of the electrical outlets in your home are connected in series or in parallel? Give at least two reasons to justify your answer.

d. Compare the one-bulb and two-bulb circuits you made. Which one do you think would cause the batteries to wear out the fastest? Why?

103

Investigation 13.2 Parallel Circuits

3 Measuring current and voltage

Now you will be taking measurements in your circuit. It is important to think carefully about how to connect the meter for each measurement.

1. Close both switches so both bulbs are on.
2. Set the meter to measure volts (DC). Connect the meter leads across the two batteries to measure their total voltage. The meter should show approximately 3 volts. Record the value in Table 1.
3. Touch one meter lead to each side of bulb 1. Record its voltage in Table 1.
4. Repeat for bulb 2.
5. Open the switch that controls both bulbs. Set the meter to measure amps (DC). Touch one meter lead to each terminal of the open switch. The bulbs should go on and the meter should display the current coming from the batteries. Record the value in Table 1.
6. Close the switch that controls both bulbs, and open the switch that controls only bulb 1. Connect the leads to the open switch to measure the bulb's current. Record.
7. Remove the switch that controls bulb 1. Place it so it controls bulb 2. Open the switch and use the meter to measure the current through bulb 2. Record the current.

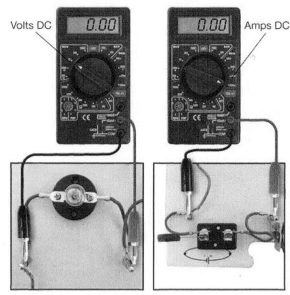

Measure voltage across the terminals of each bulb with the switch closed

Measure current across the switch terminals with the switch open

Table 1: Current and voltage values

	Batteries	Bulb 1	Bulb 2
Voltage (V)			
Current (A)			

4 Arguing from evidence

a. How does the voltage of the batteries compare to the voltage of each bulb?
b. How does the current supplied by the batteries compare to the current through each bulb?
c. Use the battery voltage, the battery current, and Ohm's law to calculate the total resistance of the circuit.
d. Use bulb 1's voltage and current to calculate its resistance. Repeat for the other bulb.
e. Is the total circuit resistance more or less than the individual resistances? Why do you think this is?

13.3 Electrical Energy and Power

How much energy is carried by electricity?

A voltage of one volt means one amp of current can do one joule of work per second. If the voltage and current in a circuit are multiplied together, the result is the power used by the circuit. In this investigation, you will use a device called a capacitor to store energy in a circuit.

Materials List
- Electric Circuits kit, including capacitor
- DataCollector or stopwatch
- 2 "D" batteries
- Multimeter with leads

1 Energy and power in an electrical system

1. Connect a simple circuit with a single bulb, switch, and battery.
2. Use the meter to measure the bulb's voltage and current in the circuit when the bulb is lit.
3. Use the formula to the right to calculate the power used by the bulb in watts.
4. Repeat the experiment with two batteries in the circuit so the bulb receives 3 V instead of 1.5 V.

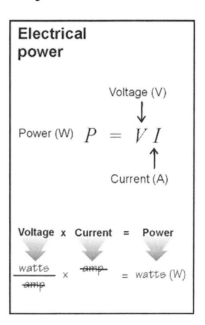

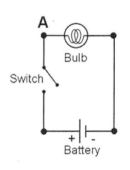

Circuit diagram

Table 1: Power used by a bulb

Number of batteries	Bulb voltage (V)	Current (A)	Power (W)
1			
2			

2 Constructing explanations

a. How did the power used by the bulb when there was one battery compare to the power when there were two batteries?

b. Was the bulb brighter, dimmer, or about the same with two batteries compared to one? Explain any difference you observed using the concept of power.

105

Investigation 13.3 Electrical Energy and Power

3 Energy and power from a battery

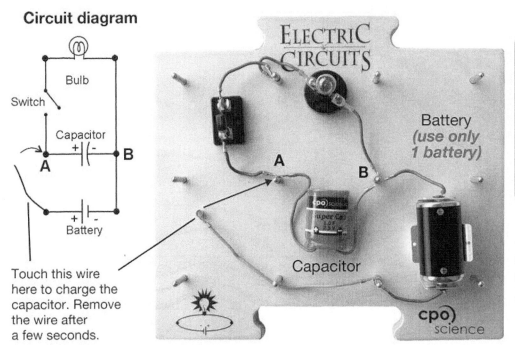

1. Find the capacitor in the circuit set. This capacitor acts like a battery that charges almost instantly when you touch its terminals to a battery.
2. Make the circuit in the diagram. The negative terminal of the capacitor should be connected to the negative terminal of the battery. The switch should be open so no current can flow through the bulb.
3. Set the multimeter to measure volts (DC). Connect the meter leads to the two ends of the capacitor. The meter should read 0 volts.
4. With the meter connected to the capacitor, touch the positive wire from the battery to the positive (+) terminal of the capacitor for five seconds. The capacitor voltage should reach a value of approximately 1.5 V. Remove the positive battery wire once the capacitor is charged.
5. Close the switch so current flows from the capacitor and through the bulb. The meter should show that the capacitor voltage drops as the bulb lights up and then dims.

4 Arguing from evidence

a. How was energy flowing when the capacitor was "charging up"? What was the source of the energy and where did it go?

b. How was energy flowing when the switch was closed and the bulb lit up? What was the source of the energy and where did it go?

c. The power of a bulb is related to its brightness. What happened to the power of the bulb as time went on? Why?

5 Energy and power

You are going to use the Stopwatch function of the DataCollector (or a simple stopwatch) to measure how long the capacitor can keep one or more bulbs lit.

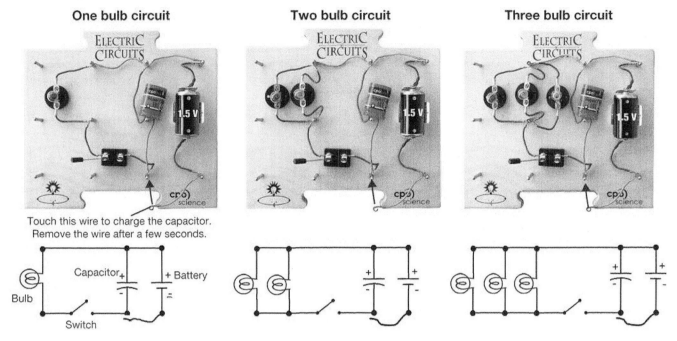

1. Set up the three circuits above, one at a time.
2. For each circuit, charge the capacitor for five seconds and then use the stopwatch to measure how long the bulb produces light. Start the stopwatch when you close the switch to light the bulb. Stop the stopwatch when you can no longer see any light.
3. Repeat the test three times and take the average. Use Table 2 to record your data.

Table 2: Energy and power data at 1.5 V

Number of bulbs	Time until bulb goes out (s)	Average of 3 trials (s)
1		
2		
3		

6 Arguing from evidence

a. How is the number of bulbs related to the time the bulbs stayed lit?

b. Which of the three circuits had the greatest power the instant the switch was closed and the bulbs went on? How do you know?

c. The energy stored in the capacitor was the same for each of the three circuits, but the powers and times differed. Use the formula *power = energy ÷ time* to explain why the bulbs stayed on for different amounts of time in the three circuits.

14.1 Electric Charge

What is static electricity?

Have you ever felt a shock when you touched a metal doorknob or removed clothes from a dryer? A tiny imbalance in either positive or negative charge on an object is the cause of *static electricity*. In this investigation, you will cause objects to have charge imbalances, and then observe what happens when different objects interact. Remember: positive and negative charges attract; like charges repel.

Materials List
- Fleece, silk, and/or fur fabric
- Plastic or rubber rod
- Styrofoam
- Aluminum foil
- Thread
- 18–22 Gauge copper wire
- Balloon
- Paper clip and pencil
- Scissors

1 Observing static electricity

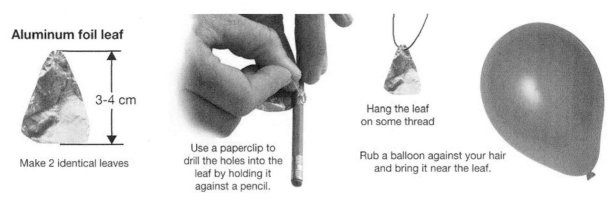

1. Cut out two small "leaves" of aluminum foil and use a paper clip to make a hole in the top of each leaf. You can hold the leaf against a pencil as you poke the hole with the paper clip.
2. Suspend one leaf from a thread that you hold up in one hand. Set the other leaf aside for later.
3. Rub an inflated balloon against your hair and move it toward the foil leaf. Observe what happens before and after the leaf touches the balloon.
4. Bring other materials, such as a plastic or rubber rod and a piece of styrofoam, near the leaf before and after they are rubbed with fur or fleece.

2 Constructing explanations

a. The rubber balloon becomes negatively charged when your rub it on your hair. What type of charge does your hair have after being rubbed on the balloon?

b. Which type of particles, protons or electrons, are transferred between your hair and the balloon? Does the balloon gain or lose these particles?

c. Describe what happens to the aluminum foil leaf as you move the balloon closer.

d. Describe what happens to the aluminum foil leaf after it touches the balloon. Explain your observation using the concept of positive and negative charge.

e. Describe the observations you made when bringing other materials toward the foil leaf.

3 Building an electroscope

1. Cut a piece of insulated copper wire so it is about 10 centimeters long.
2. Strip about 2 centimeters of the insulation from the wire at both ends. Be careful not to break the wire.
3. Bend the wire over the edge of the glass beaker as shown in the photo.
4. Make the electroscope as shown in the photo by hanging two leaves on the end of the wire that is inside the beaker.
5. Rub an inflated balloon against your hair and move it toward the end of the wire sticking out of the beaker.
6. Touch the balloon to the wire, and then remove it.
7. Touch the end of the wire with your finger or a metal object.
8. Repeat steps 5 through 7 with other materials, such as styrofoam and fur.

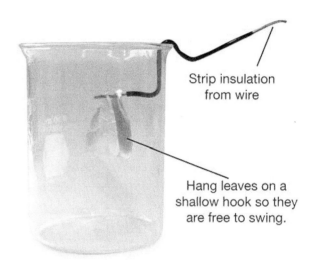

Strip insulation from wire

Hang leaves on a shallow hook so they are free to swing.

4 Arguing from evidence

a. Describe what happens to the aluminum foil leaves as you move the rubbed balloon closer to the wire.

b. Give a reason why the leaves stay apart after the balloon is touched to the wire and then is removed. Use the concept of electric charge in your answer.

c. Explain what happens when you touch the wire with your hand or a metal object.

d. Describe what you observed when you used other materials such as styrofoam, fur, and fleece with the electroscope.

14.2 The Flow of Electric Charge

How much charge moves when current flows?

One electron is a very tiny amount of charge. The flow of current represents many, many electrons moving together. This investigation will lead you inside the submicroscopic world of the atom to deduce how many electrons really move when current is flowing in a circuit.

Materials List
- Electric Circuits kit including capacitor
- Multimeter with leads
- "D" battery
- Graph paper
- Calculator

1 Measuring voltage

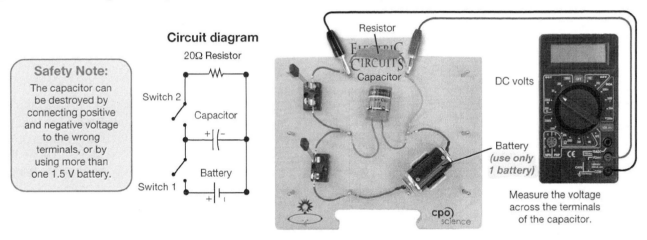

Safety Note: The capacitor can be destroyed by connecting positive and negative voltage to the wrong terminals, or by using more than one 1.5 V battery.

1. Find the resistor with a resistance of 20Ω.
2. Connect the circuit shown using the battery, the capacitor, the resistor, and two switches.
3. Attach the meter to measure the voltage across the resistor.
4. Charge the capacitor by closing switch 1. This makes a connection between the positive wire from the battery and the positive (red) lead of the capacitor. Count five seconds and then open switch 1.
5. Make sure switch 1 is open. Close switch 2 and record the voltage across the resistor every ten seconds in Table 1. You may stop recording when the voltage drops to below 0.05 volts.

Table 1: Capacitor discharge data

Time (s)	Voltage (V)	Time (s)	Voltage (V)
0		70	
10		80	
20		90	
30		100	
40		110	
50		120	
60		130	

2 How much current flowed?

a. Use Ohm's law to write down a formula for the current flowing through an electrical device if you know the voltage and the resistance.

b. Use the relationship you found in question 2a to fill in Table 2 by calculating how much current was flowing through the resistor.

Table 2: Capacitor discharge data

Time (s)	Current (A)	Time (s)	Current (A)
0		70	
10		80	
20		90	
30		100	
40		110	
50		120	
60		130	

3 Analyzing your data

a. Make a graph of the current versus time for the capacitor. Draw a smooth curve through the points on your graph.

b. Describe what your graph shows about the current over time. Explain the reason for the current to change in this way.

c. Suppose a light bulb had been used instead of a resistor. What would you have noticed about the light bulb as time went on? Why?

d. Calculate the power of the resistor at a time of ten seconds and at a time of 60 seconds. Do your answers support your explanation in question 3c?

4 Calculating charge from your graph

a. Write down a formula that allows you to calculate charge if you know the current and the time.

b. If current is in amperes and time is in seconds, what is the unit of charge represented by your formula from question 4a?

Investigation 14.2 The Flow of Electric Charge

c. Look at the scale on your graph and determine the time and current represented by one block. Draw one graph block and label its dimensions. See the example to the right.

d. Use your answers to questions 4a and 4b to calculate the amount of charge represented by one block on your graph. Show your work.

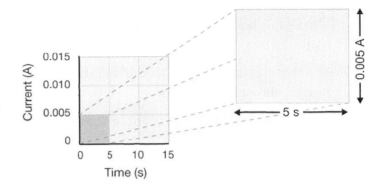

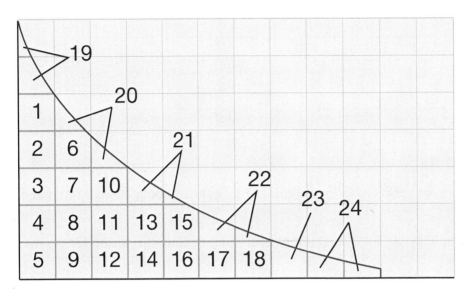

Count the total number of grid blocks under the curve. Add partial blocks together to make whole blocks.

e. Count the number of blocks between the curved line on your graph and the *x*-axis as shown above. Include fractions of blocks in your count, and estimate as accurately as you can.

f. In question 4d, you figured out the amount of charge represented by one block. Use this information and the number of total blocks under your graph to calculate the total charge that flowed from the capacitor and through the circuit.

g. One coulomb is equal to 6.24151×10^{18} electrons. Calculate the number of electrons that flowed from the capacitor.

h. Explain how the results of your investigation would differ if you used a 30Ω resistor instead of a 20Ω resistor. If you have time, you may want to try the experiment with a resistance of 30Ω (20Ω and 10Ω resistors in series) to see if you are correct.

14.3 Making an Electrophorus

How do electric charges interact?

If you have ever felt a shock when touching a doorknob or removing clothes from the dryer, you have experienced the effect of electric charge. Electric charge is a fundamental property of matter. All ordinary matter contains both positive and negative charge. You do not usually notice the charge because most matter has equal amounts of both types of charge. Interesting things can happen when objects can lose or gain electric charges, as you will see in this investigation.

In this investigation, you will:

- triboelectrically charge different materials.
- use a triboelectric series to make predictions about charged objects.
- make an electrophorus and explain how it works.

Materials List

- Static electricity box
- Piece of wool
- Modeling clay
- PVC tube (from wave tray)
- Aluminum block (from wave tray)

1 Experimenting with an electrophorus

An electrophorus is a device invented over 200 years ago. A simple electrophorus has a charging surface and a metal plate with an insulating handle. The metal plate becomes charged and you can carry the charge to various places to perform static electricity experiments. Follow the steps below to make an electrophorus, and then use the triboelectric series and your knowledge of how charges interact to explain how the electrophorus works.

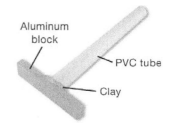

1. Attach the PVC tube to the aluminum block with some modeling clay (see photo at right).

2. Vigorously rub a piece of wool over the upside-down static electricity box at least 100 times. This is the charging surface for the electrophorus.

3. Hold the PVC tube handle and place the aluminum block down on the area of the static electricity box that you rubbed with the piece of wool. Touch a finger to the aluminum block next to the handle (see photo at right). Remove your finger from the aluminum. Pick up the aluminum block by the handle.

4. Hold your knuckle near the aluminum block and observe what happens. If you can do this in a dark area of the room, it will be more dramatic.

5. Repeat step 3. Use the aluminum block to conduct static electricity experiments of your own. Each time you discharge the aluminum block, you can recharge it by repeating step 3 (you won't have to repeat step 2 every time).

Investigation 14.3 Making an Electrophorus

2 Arguing from evidence

a. Use the triboelectric series to figure out the type of charge the acrylic static electricity box has after being rubbed with the wool.

b. What is the function of the PVC tube on the electrophorus? Why can't you just pick up the aluminum block with your hand and move it around?

c. Aluminum is a good conductor of electrons. When the aluminum block is first placed on the charged acrylic static electricity box, the electrons tend to move to one area of the aluminum block, leaving one area positively charged and one area negatively charged. When you touch the aluminum block with your finger, the electrons will follow the conductive path to your finger and *leave the block*. With this information, complete the following sequential explanation of how the electrophorus works.

- When the aluminum block is placed on the charged acrylic static electricity box, the electrons move to the _____ (top or bottom) of the block, leaving the _____ (top or bottom) of the block positively charged.

- Acrylic is an insulator, so charges do not easily move between the acrylic and the aluminum block. At this point, the aluminum block is still neutral, since no charge has entered or left the block.

- When you place your finger on the aluminum block, the _____ (electrons or protons) are conducted away from the block to your finger, leaving the aluminum block with a _____ (positive or negative) charge.

- When you bring the charged aluminum block near a conductor like a piece of metal or one of your fingers, electrons move from _____ to the _____, causing the aluminum block to become neutral again.

d. Why can you recharge the aluminum block on the charged acrylic static electricity box without having to rub wool over the box tray again?

e. Why does the electrophorus work better in a dry environment than it does in a humid environment?

f. Illustrate how an electrophorus works, step by step. Show charged areas of objects.

Triboelectric series

Most Positive
air
human hands, skin
glass
human hair
wool
silk
aluminum
paper
steel
wood
balloon
copper
acrylic
cellophane tape
polyvinyl chloride (PVC)
Most Negative

15.1 Magnetism

How do magnets and compasses work?

Magnets are used in almost all electrical and electronic machines, from motors to computers. How far does magnetic force reach? How can you use a compass to detect magnetic forces? In this investigation, you will use magnets and a compass to answer these and other questions about magnetism.

Materials List
- 2 Magnets (from kit)
- Compass (from kit)
- Metric ruler

1 How far does magnetic force reach?

How far does the magnetic force of a magnet reach? This is an important question that affects machines such as motors and generators that use magnets.

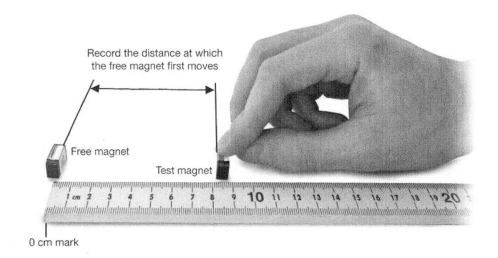

1. Place one magnet at the 0 cm mark of the ruler and slowly slide a second magnet closer until the first magnet moves. Practice the technique several times before recording data.
2. Record the distance between the magnets when you first see movement.
3. Try each of the combinations of poles—north-north, south-south, and north-south.
4. For each combination, complete three trials, and average your three distances.

Table 1: Magnetic forces between two magnets

	North-south	South-south	North-north
Distance 1 (mm)			
Distance 2 (mm)			
Distance 3 (mm)			
Average distance (mm)			
Average estimated error (mm)			

Investigation 15.1 Magnetism

2 Constructing explanations

a. What is the average estimated error for each magnet combination? Subtract each individual distance for the north-south magnet combination from the average of the three distances. Drop any negative signs. Once you have found these three differences, average them and record in Table 1. Repeat for the other two magnet combinations.

b. Are the attract and repel distances *significantly* different? Your answer should include a comparison between average estimated errors and the differences between magnet combination average distances.

3 Using a compass to detect magnetic forces

The needle of a compass is a permanent magnet. Earth is magnetic, so a compass needle is attracted to north in the absence of other (stronger) magnets.

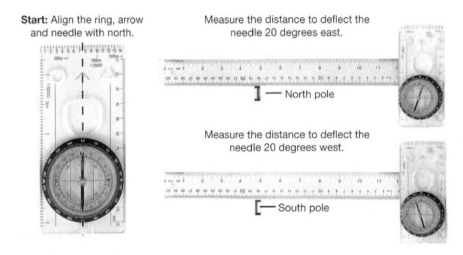

1. Set a compass on your table far from any magnets. Rotate the compass so the needle, dial, and arrow are all aligned with north.
2. Place a metric ruler to the side of the compass and line it up to be perpendicular to the north pole of the compass. Move a small magnet near the compass and note the distance at which the needle moves 20 degrees from north.
3. Reverse the pole of the small magnet and note the distance at which the needle moves 20 degrees in the opposite direction.

4 Arguing from evidence

a. At a distance of 10 cm, which is stronger: the magnetic force from Earth, or the magnetic force from the small magnet? How is your answer supported by your observations?

b. Is the end of the compass needle a magnetic north or a magnetic south pole? How is your answer supported by your observations?

c. Is the geographic north pole of the planet Earth a magnetic north or a magnetic south pole? How is your answer supported by your observations?

15.2 Electromagnets

How are electricity and magnetism related?

Almost every electrical device that creates motion, such as a motor, uses magnets. Permanent magnets are not the only magnets used in these devices. Often, electromagnets are used. Electromagnets create magnetic forces through electric currents. This investigation will explore the properties of electromagnets.

Materials List
- Electric Circuits kit
- Electromagnet coil (in kit)
- Permanent magnet (in kit)
- Compass (in kit)
- "D" battery

1 Electromagnets

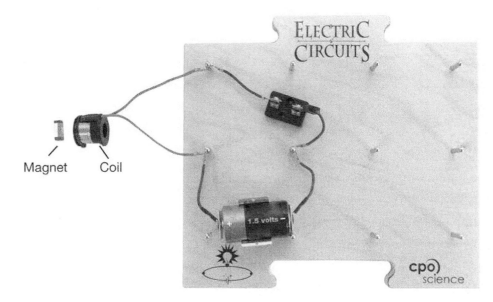

1. Attach the coil, battery, and switch in the circuit shown above. Leave the switch open so no current flows.
2. Place a permanent magnet about 1 centimeter away from the coil. Stand the magnet up on its end.
3. Close the switch and watch what happens to the magnet. *Do not* leave current running or the coil will overheat. Open the switch after each trial.
4. Turn the permanent magnet around so its other pole faces the coil. Close the switch and see what happens now.
5. Reverse the wires connecting the battery to the circuit. This makes the electric current flow the other way. Repeat steps 3 and 4 above.

2 Constructing explanations

a. Write two to three sentences that explain what you saw when the switch was closed.

b. Propose an explanation for why the magnet moved.

c. When the magnet was reversed, did the force between it and the coil change direction? How did the force change?

Investigation 15.2 Electromagnets

d. When the coil wires were switched, did the force from the coil change direction? How do you know?

e. How is a current-carrying coil like a magnet? How is it different? Explain how this shows that electricity and magnetism are related.

3 Comparing the electromagnet to a permanent magnet

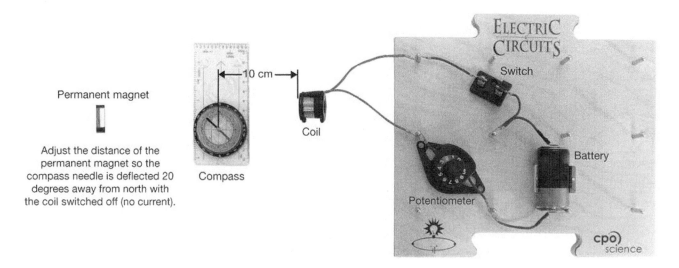

1. Attach the potentiometer, coil, battery, and switch in the circuit shown in the diagram. Leave the switch open so no current flows.
2. Set the compass so the needle, ring, and arrow are all aligned with north. Put the coil about 10 cm from the center of the compass.
3. Place a permanent magnet on the side of the compass opposite the coil. Bring the magnet close enough to deflect the needle 20 degrees away from north.
4. Close the switch and adjust the potentiometer so the needle returns to north. The coil should deflect the compass needle back toward north. Reverse the permanent magnet if the needle moves the wrong way. *Do not* leave current running or the coil will overheat. Open the switch after each trial.
5. Try moving the permanent magnet to different distances, along with using the potentiometer, to return the compass needle to north with force from the electromagnet.

4 Analyzing and interpreting evidence

a. The permanent magnet is pulling the compass needle to the left. The electromagnet is pulling the needle in the opposite direction to the right. When the needle returns to north, what can you say about the magnetic forces from the permanent magnet and electromagnet?

5 Iron and electromagnets

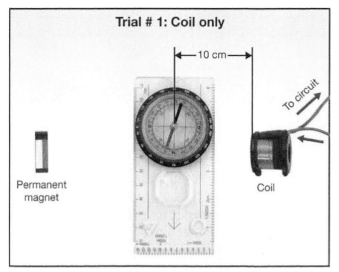
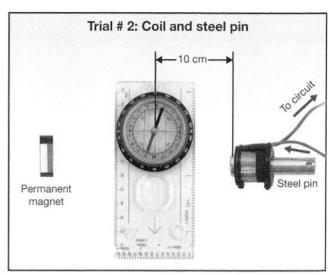

Measure the current it takes to bring the compass needle back to north.

1. Use the same circuit as for Part 3 with one battery and the switch, coil and potentiometer.
2. Rotate the compass until the needle and dial are aligned with north. *There should be no magnets nearby, and no current in the coil for this step.*
3. Move a permanent magnet close enough to deflect the needle 20 degrees from north.
4. The coil should be 10 centimeters from the center of the compass (see diagram above). Close the switch, then use a multimeter to measure and record how much current it takes for the coil to bring the needle back to north. Adjust the current with the potentiometer. Once you have recorded the measurement in Table 1, open the switch to stop the current.
5. Put the steel pin in the coil so its head is against the coil and 10 centimeters from the center of the compass.
6. Adjust the distance of the permanent magnet so the compass needle is deflected 20 degrees from north, like you did in step 3. *There should be no current in the coil for this step.*
7. Close the switch, then use a multimeter to measure and record the current it takes to return the needle to north with the steel pin in the coil.

Table 1: Electromagnet current with and without the steel pin

Current with bare coil	Current with steel pin	Difference in current	Percent difference

6 Arguing from evidence

a. How did the steel pin affect the magnetic force created by the coil? Was the magnetic force reduced or increased, or did it stay about the same? Use your observations to support your answer.

15.3 Making a Model Maglev Train

How can you make a model maglev train?

A magnetically levitating (maglev) train does not roll on wheels. Instead, it uses electromagnets to lift the train so it hovers above the tracks. Maglev technology is still in its experimental stages. However, many engineers believe that maglev trains will be used worldwide within the next 100 years.

In this investigation, you will research maglev trains and create a model maglev train.

Materials List

- 30 or more magnets with north and south poles on the faces, rather than on the ends
- Model-building supplies, such as cardboard, foam core, dowel rods, paper, tape, glue, and the like
- Materials for propelling the train, such as a balloon, rubber band, or similar
- Log book (notebook or stapled pages)

1 Researching maglev trains

Use the Internet to research maglev trains and find answers to the questions below. Organize your answers into an essay or create a poster.

a. What are the main parts of a maglev train?
b. How are maglev trains propelled?
c. What are the benefits of using maglev trains?
d. Where are maglev trains currently in use? Where are they being built for use in the near future?

2 The engineering cycle

You will use permanent magnets to design a model maglev train of your own. When engineers design technology such as cars, computers, artificial hearts, or maglev trains, they follow a process called the engineering cycle. The parts of the cycle are:

- Evaluate: Define the problem and set the goal of the project. Identify the constraints and variables.
- Design: Brainstorm to create a list of ideas for the project. Be creative. It is okay if some ideas seem far-fetched. Do some research to refine your ideas. Then, select the best ideas. Use the ideas to create a design. Choose materials, make drawings, and decide what you will build.
- Prototype: Follow your design to build a prototype. Keep a record of any difficulties you have while building.
- Test: Find out if your prototype works. Make notes about its strengths and weaknesses.

Once the prototype is tested, return to the "evaluate" step and brainstorm new ideas for improving the design. The engineering cycle is repeated as many times as needed to create and improve products.

3 Using the engineering cycle

Keep a log book to document how you use the engineering cycle to make a model maglev train. Your log book should contain all of your notes, drawings, observations, and testing results. The guidelines below will help you work through the engineering cycle.

Evaluate

Make sure you are clear about the goals and constraints of the project. Think about how the following constraints will affect the type of maglev train model you can build.

- Your train must use permanent magnets to levitate. Use the magnets supplied by your teacher.
- Use any materials you wish for the body of the train and for the track. Common materials are cardboard, paper, plastic containers, wrapping paper rolls, and wooden dowels.
- Your train must move at least 50 centimeters along the track after it is released from rest.
- The propulsion device can be part of the train car, the track, or both.
- Trains will be judged on three criteria: time it takes to travel 50 centimeters, design creativity, and quality of construction.

Design

1. Brainstorm ideas for your train. Make notes and sketches of how you might build the train car and the track. List possible materials you can use.
2. Brainstorm ideas for the train's propulsion. How will your train travel 50 centimeters?
3. Study your ideas. Select the best design and method of propulsion. Make a drawing of your design. Include measurements and label the materials.

Prototype

1. Build your prototype.
2. Make notes in your log book about any changes you make to the original design.

Test

1. Test your prototype. Make notes about what worked well and what didn't work.
2. Go back to the "evaluate" step to improve your model. You may choose to use a completely different design, or you may want to think about ways to refine your current design. Keep going through the cycle until you have the best train you can build.

4 Comparing and evaluating models

How do your classmates' train designs compare to yours? Make comments about what you like about each design and what could be changed to make it better. Here are some guidelines to help you rate each model.

1. Watch the maglev model train in action. Record the time it takes to travel 50 centimeters.
2. Rate the train on a scale of 1 to 10 for design creativity and quality of construction.
3. Make comments about the strong points of the design and the ways the train could be improved.

16.1 Electromagnetic Forces

How does an electric motor work?

Electric motors are found in many household devices, such as hair dryers, blenders, drills, and fans. In this investigation, you will build a simple electric motor and see how it works. The concepts you learn from the simple motor also apply to other electric motors.

Materials List

- Permanent magnet (from kit)
- "D" battery
- Large rubber band
- Metric ruler
- Half-stick of modeling clay
- Sandpaper
- Varnished magnet wire
- Paper clips
- Photogate (optional)
- DataCollector (optional)

1 Making the base

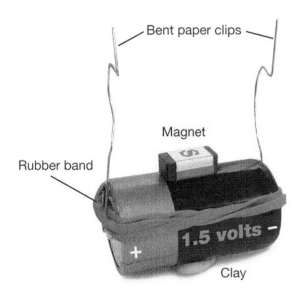

1. Bend the two paper clips so they look like the photo.
2. Fasten them to the battery with rubber bands so they contact the positive and negative terminals.
3. Break off a small lump of clay from the half-stick. Set the battery on the small clay lump so it stays in one place without rolling around.
4. Use a tiny piece of clay to stick a magnet to the top of the battery.
5. Your motor base is complete!

Electromagnetic Forces   Investigation 16.1

2 Making the coil

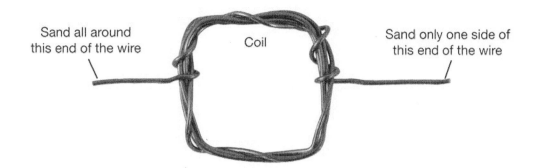

1. Cut one meter of magnet wire (also called varnished magnet wire). This wire has a painted insulation layer on the surface.
2. Wrap the wire around the square form of the modeling clay with one end sticking out about four centimeters.
3. Keep wrapping until you have only four to five centimeters left.
4. Remove your coil from the form and wrap the ends of the wire a few turns around the sides of the coil to keep things together. There should be about three centimeters of wire on each side of the coil.
5. Take some sandpaper and sand all the varnish off one end of the wire.
6. Sand off the varnish *on one side only* of the other end of the wire.
7. Adjust the wires until the coil balances as well as you can get it.
8. Your coil is done!

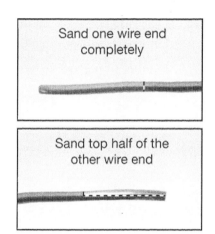

3 Making the motor work

1. Set the coil into the paper clips so it is free to spin.
2. Adjust the height of the paper clips until the coil rotates just above the magnet.
3. The motor should spin! Adjust the balance by bending the wires or paper clips.

Troubleshooting—The main problem areas to watch out for are:

- unbalanced coil
- poor connections
- improperly sanded coil ends
- dead D-cell

123

Investigation 16.1 Electromagnetic Forces

4 Constructing explanations

a. When electricity runs through the coil of wire, what type of force acts on the coil?

b. What is the purpose of the permanent magnet?

c. What interactions cause the coil of wire to spin?

d. Try adding a second magnet. Does this make the motor go faster, slower, or about the same? What observations did you make that support your answer?

e. What else might make the motor spin faster? With your teacher's approval, try it!

5 Planning and carrying out investigations

See if you can hold a photogate so the spinning coil breaks the light beam. With the DataCollector's Timer mode set on Frequency, you will know how many breaks per second the spinning coil makes.

a. Describe how you set up the investigation to get frequency readings from the DataCollector/photogate equipment.

b. How did your spinning coil measurements compare to other groups' data?

c. Repeat the last two steps in Part 4 (4d and 4e). Record your data.

d. Does the spinning coil data support your answers to 4d and 4e? Explain.

e. The basic parts of any simple DC motor like the one you made are an electromagnet, permanent magnet, commutator, and energy source. Draw a diagram of your simple motor and label these parts.

16.2 Electromagnetic Induction

How does an electric generator work?

Changing electric currents can cause magnets to move, as in the electric motor you built. The reverse is also true: changing magnetism can cause electric currents to flow. This is the principle on which the electric generator works. In this investigation you will build a simple electric generator and test its voltage.

Materials List
- Magnets
- Electromagnet coil
- Ripcord Generator kit
- Digital meter and leads
- Electric Circuits kit
- Safety goggles
- DataCollector
- Graph paper
- Photogate

  (1) Wear safety goggles, (2) pull the ripcord with a straight motion, pulling directly away from the spindle, (3) firmly hold the base of the generator assembly and circuit board.

1 Making a generator

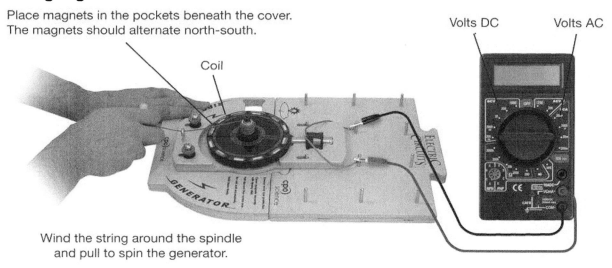

1. Attach the Ripcord Generator assembly to the circuit board.
2. Connect the meter and coil as shown in the diagram.
3. Put four magnets in the rotor, evenly spaced and alternating north/south.
4. Wrap a string around the spindle and pull it to set the rotor turning.
5. Observe the voltage with the meter set first to "Volts DC" and then "Volts AC."

2 Constructing explanations

a. Explain the difference between AC and DC electricity.

b. Is electricity produced by the magnets at rest or only by motion of the magnets?

c. Does the generator make AC or DC electricity? Support your answer with your observations.

Investigation 16.2 Electromagnetic Induction

3 Voltage and speed

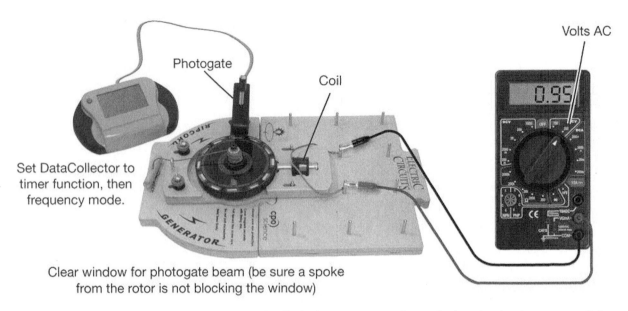

1. Slip a photogate under the rotor so the light beam passes through the slot in the cover of the rotor. Use the DataCollector in Timer mode and set it to measure Frequency (f). This tells you how many times per second the light beam is broken. The beam is broken once per turn.
2. Start the rotor spinning. As it slows down, record the AC voltage produced by the generator at different speeds. This will take several people cooperating. You will need to develop and practice a technique for recording numbers that are changing rapidly.

Table I: Voltage versus speed data

Frequency from Timer (Hz)	Rotational speed (rev. per s)	Voltage produced (V AC)

4 Analyzing your results

a. Does the voltage produced depend on the speed? Support your answer with your observations.

b. Make a graph of voltage versus speed. Is the graph a straight line or a curve?

5 Changing the design

1. Use different numbers of magnets.
2. Adjust the difference between the coil and the rotor.

6 Arguing from evidence

a. Which of the changes that you made in Part 5 have the greatest effect on the voltage produced?

b. Explain why one change made a large difference while the other did not. The answer is not obvious, so this should be a class discussion question.

7 Building different generators

Now you are going to test different configurations of magnets at different speeds.

6 magnets - alternating

12 magnets - alternating

12 magnets - same side out

1. Try different combinations of magnets in the rotor.
2. Try facing all the magnets the same way.
3. Measure the voltage at a constant speed for each different magnet configuration.

Table 2: Voltage for different magnet configurations

Magnet configuration (describe)	Rotational frequency (Hz)	Voltage produced (V AC)

8 Constructing explanations

a. If the number of magnets is increased from six to 12, what change do you expect in the voltage? Assume the magnets alternate north-south.

b. Suppose you have six magnets (alternating north-south) and double the speed. What change do you expect in the voltage?

c. Suppose you change from six to 12 magnets and also double the speed. What change do you expect in the voltage?

d. Propose a relationship that accounts for the voltage produced at different speeds and magnet configurations. This should be a class discussion question.

e. What change in voltage do you expect when changing from 12 magnets alternating north-south to 12 magnets all facing the same way? Why?

16.3 Generators and Transformers

How do electricity and magnetism work together in generators and transformers?

In the early 1800s, English physicist Michael Faraday (1821-1867) discovered that moving a magnet through or near a loop of wire creates electric voltage and current in the wire. This relationship between electricity and magnetism is used to generate most of the electricity used today. This investigation will give you a better understanding of generators and transformers, two devices that demonstrate the way electricity and magnetism are connected.

Materials List
- Computer with internet access
- Faraday's Electromagnetic Lab PhET simulation
- DataCollector (power adapter only)

1 Creating current with a simple generator

In the previous investigation, you measured the voltage created by using the Ripcord Generator to produce electricity by moving some permanent magnets past a stationary coil. In this investigation, you will use a simulation to examine how current can be produced in a similar manner.

a. Open the Faraday's Electromagnetic Lab computer simulation at this address: http://phet.colorado.edu/en/simulation/faraday.

b. Click on the "Pickup Coil" tab. Move the magnet and describe what happens.

c. What are the blue particles in the wire that move when there is current in the wire?

d. Move the magnet through the coil of wire at different speeds. How is the brightness of the bulb related to the speed?

e. Select the voltmeter in the indicator box. Move the magnet through the loop and describe what the voltmeter shows.

f. A simple generator is a coil of wire with a magnet moving through it. Experiment with the generator and describe three ways its design can be changed.

2 A more complex generator

a. Click on the "Generator" tab on the simulation. This shows you how a waterfall can be used to make electricity. Slide the button on the faucet handle to turn the water on. Describe what happens.

b. Does this generator create direct or alternating current?

c. Select the voltmeter in the indicator box. Describe two ways to increase the voltage created by the generator.

Generators and Transformers   Investigation 16.3

3 Using an electromagnet

a. Click on the "Transformer" tab in the simulation. This simulation is just like the pickup coil simulation, but it uses an electromagnet instead of a bar magnet. Describe two ways you can change the strength of the electromagnet.

b. Adjust the electromagnet so it has four loops and 10 volts. Adjust the pickup coil so it has three loops and a loop area of 100%. Place the electromagnet inside the pickup coil. Move the slider on the battery back and forth. Describe what you observe.

c. Select "AC" under the current source. Describe what you observe.

4 Using a transformer

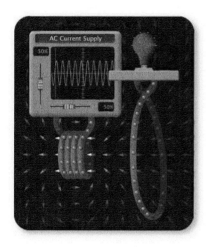

A transformer is a device that contains two coils. The *primary coil* is connected to the source of the electricity, such as the electrical line that goes to a power plant. The changing magnetic field created by the primary coil creates voltage and current in the *secondary coil*. For a transformer on an electrical pole, the secondary coil connects to wires that go to houses and other buildings. The amount of voltage in the secondary coil can be changed by changing the number of loops in the two coils.

a. Move the coils so they are side-by-side as shown at left. Set the pickup coil to have one loop. Which is the primary coil, and which is the secondary coil?

b. Adjust the two sliders on the AC current supply and describe what they do.

c. Set both of the sliders to approximately 50 percent. In which coil are the electrons moving more?

d. Move the coils so one is inside the other as shown at right. Describe what happens to the motion of the electrons in the pickup coil.

e. Increase the number of loops in the pickup coil to three. Describe what happens to the motion of the electrons.

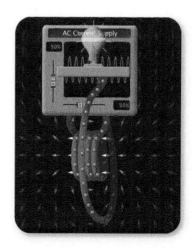

Investigation 16.3 Generators and Transformers

5 Transformer coils and voltage

You saw that by changing the number of loops in the primary and secondary coils of the transformer, you can change the voltage inducted in the secondary coil. When analyzing transformers, loops are often called "turns" because what really matters is the number of times the wire "turns" around the core. There is a mathematical relationship between the number of turns and the voltage induced.

The transformer in the graphic has both coils wrapped around a common iron core. Many of the transformers you use every day utilize an iron core, like the wall adapter used to power the DataCollector.

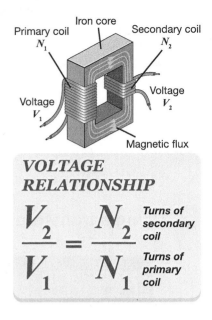

VOLTAGE RELATIONSHIP

$$\frac{V_2}{V_1} = \frac{N_2}{N_1}$$

N_2 Turns of secondary coil
N_1 Turns of primary coil

a. Look for the list of the input voltage and output voltage on the side of the DataCollector's wall adapter. What is the input voltage? What is the output voltage?

b. Which voltage is the voltage of the primary coil?

c. Which voltage is the voltage of the secondary coil?

d. Using the voltage relationship given in the graphic to the right, if the primary coil has 1,000 turns, how many turns does the secondary coil have?

e. Using a primary coil with 1,000 turns, what would the output voltage be if the secondary coil has 50 turns?

17.1 The Magnetic Field

Does magnetic force spread out in the area around a permanent magnet?

How does one magnet "know" another magnet is there? The answer took a long time to discover, and has shaped much of our understanding of physics. This investigation will explore the concept of a field using the magnetic field as an example. Almost all of what you learn applies to other fields we know to exist, such as gravitational fields and electric fields.

Materials List
- Magnets
- Tape
- Compass
- Cardboard, 50 cm x 70 cm or larger

1 Making a map of a magnetic field

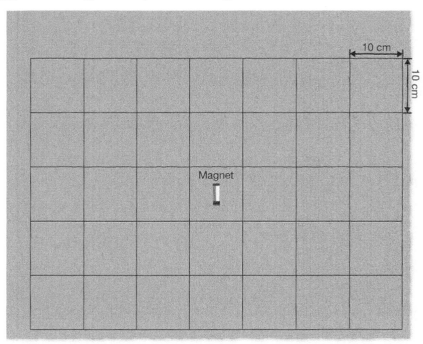

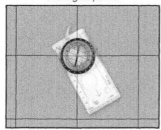

Put the center of the needle on each grid point.

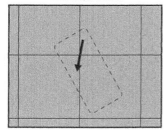

Remove the compass and draw an arrow in the direction the needle pointed.

1. Draw and label *x*- and *y*-axes on two perpendicular edges of the cardboard. Draw a grid of 10-centimeter boxes.
2. Tape the board to a table so it cannot move.
3. Tape a magnet to the center of the cardboard with the north-south axis parallel to the long side of the cardboard.
4. Put the compass on a grid point so the center of the needle is over the grid point.
5. Observe the angle of the needle, then remove the compass and draw an arrow in the direction the north (colored) end of the needle was facing.
6. Map out the whole board, one grid point at a time.

Investigation 17.1 The Magnetic Field

2 Constructing explanations

a. The arrow of the compass shows you the direction of the force felt by the _____ pole of another magnet. (Fill in the blank with "north" or "south.")

b. The lines of magnetic field point away from _____ and toward _____ magnetic poles. (Fill in the blanks with "north" or "south.")

c. Why do all the arrows that are far from the magnet point in the same direction?

d. Research and give a one-sentence definition of a *field* as the term is used in physics. (This may be a class discussion question.)

e. Describe the two basic kinds of fields and give the difference between them. The magnetic force around a magnet is an example of one and the temperature around a candle flame is an example of the other.

3 The electric field

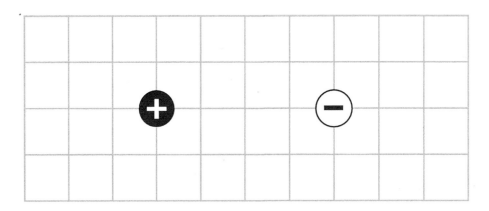

a. Draw arrows on each of the grid points in the diagram above to show the force a positive charge would feel if it were at that location. This is a diagram of the electric field.

4 The gravitational field

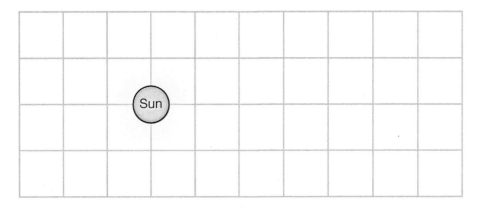

a. Draw arrows on each of the grid points in the diagram above to show the force an asteroid would feel if it were at that location. This is a diagram of the gravitational field.

17.2 Using Fields

What does a field tell you about what created it?

There is a very important law in physics stating that if you can measure every detail of a field, you can deduce everything about the object that created it (the source). In many cases, the field is what we "see," but we want to know about the object that created the field. In this Investigation, you will use magnetic fields to solve a puzzle, demonstrating how fields are related to the objects that create them.

Materials List

- Magnets
- Tape
- Compass
- Meter stick
- Cardboard, 60 cm x 90 cm

1 Making the puzzle

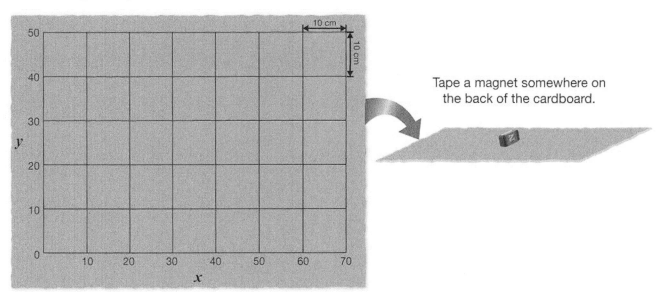

1. Draw and label *x*- and *y*-axes on two perpendicular edges of the cardboard. Draw and label the grid with 10-centimeter boxes.
2. Tape a magnet to the opposite side of the cardboard with its north-south axes parallel to the surface. Do not show any other groups where you have placed your magnet.
3. The magnet may be anywhere on the cardboard, but leave 10 centimeters of space around the edge.
4. Measure and write down the (*x*, *y*) coordinates of your magnet, and draw a diagram showing the exact angle of the north-south axis.

2 Thinking about what you are going to do

a. Exchange your puzzle with another group.
b. Locate the magnet on the cardboard grid you receive and deduce its north-south angle using only a compass to detect the field created by the magnet.

Investigation 17.2 Using Fields

3 Solving the puzzle

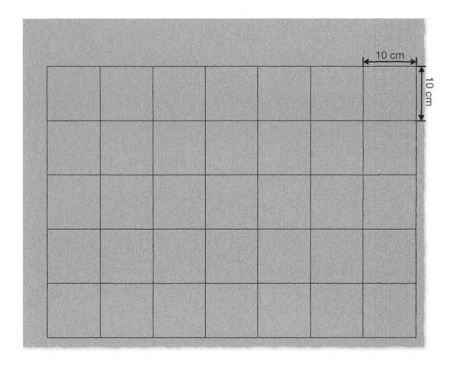
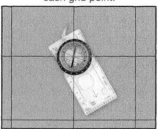
Put the center of the needle on each grid point.

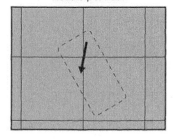
Remove the compass and draw an arrow in the direction the needle pointed.

1. Put the compass on each of the grid points so the center of the needle is over the grid point.
2. Observe the angle of the needle, then remove the compass and draw an arrow pointing in the same direction the north end of the needle was facing.
3. Map out the whole board.

4 Using your model

a. How can you use the map you made to determine the location and orientation of the hidden magnet? The diagrams below may assist you to interpret the field patterns you drew.

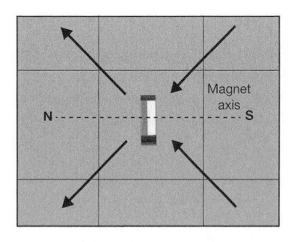
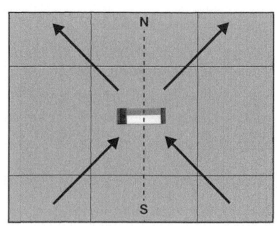

b. Sketch the magnetic axis of the magnet on your map.
c. Flip the cardboard grid over and check your results. Did you get the location and orientation correct?

17.3 Electric Forces and Fields

What are electric fields?

A field is a way for one object to exert a force on another without the two ever touching. Earth has a gravitational field that allows it to pull dropped objects down to its surface. Positive and negative charges have electric fields around them. In this investigation, you will learn how to interpret drawings of electric fields and how to calculate the force between two electrically charged particles.

Materials List
- Computer with internet access
- Charges and Fields PhET simulation
- Calculator

1 Using the simulation

1. Open the Charges and Fields computer simulation with this address: http://phet.colorado.edu/sims/charges-and-fields/charges-and-fields_en.html
2. You will see a blank playing board on which you can place charges.
3. Select one of the positive charges and place it in the middle of the board.
4. Select one of the E-field sensors. The sensor is a *test charge* that feels the force of the positive charge in the middle of the board.
5. A vector shows the force on the sensor. Move the sensor around the board and observe what happens to the vector.

2 Using your model

a. What happens to the length of the vector as you move the sensor closer to the positive charge?

b. What does your answer to the previous question tell you about the strength of the force on the sensor as it moves toward the charge?

c. Is the sensor a positive or negative charge? How do you know this?

3 Drawing the electric field

a. Use the computer simulation to move the sensor around the positive charge to all of the different locations in the diagram. Draw the force vector at each position.

b. Remove the positive charge and replace it with a negative charge. Draw the force vectors on the diagram below.

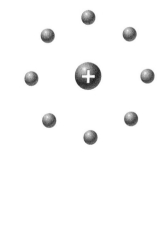

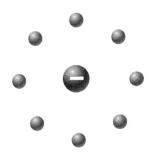

c. The drawings you have created show the *electric field* around a positive charge and around a negative charge. Do electric fields point toward positive charges or negative charges?

Investigation 17.3 Electric Forces and Fields

d. The vectors you drew are called *electric field lines*. Are the electric field lines closer together or farther apart in locations where the force on the sensor is strongest?

4 More complex electric fields

When two or more charges are in a region, the electric field diagram becomes more complicated. Instead of drawing individual force vectors, electric field lines are made by joining the vectors into curves.

a. Use the simulation to determine whether each of the charges in the diagrams below is positive or negative. You can use the field sensor or check the "Show E-field" box to display the field.

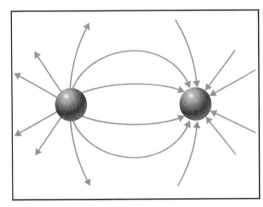 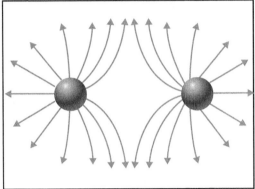

b. Write a short paragraph to summarize what you have learned about electric fields during this activity.

c. Place three charges on the screen. Check the "Show E-field" box to display the field. Label your charges with their correct charge and sketch the field you see around the charges as displayed by the vector arrows on the screen.

5 Calculating force in an electric field

Charges are often associated with objects or particles. Many useful scientific tools use accelerated particles to accomplish tasks such as displaying images on display screens and examining objects and surfaces with electron microscopes. Other times, particles can be used to examine objects or surfaces, like when using an electron microscope. To do these things, charged particles need to move at a particular speed and direction. One way to do this is with charged metal plates and a screen with holes in it.

The diagram at right shows a device that was made to accelerate electrons. The field strength in this device is 10,000 V/m.

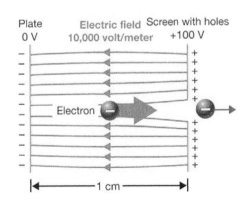

a. The charge of an electron is 1.602 x 10⁻¹⁹ coulombs. Use the electric force formula at right to calculate the force on the electron in this electric field.

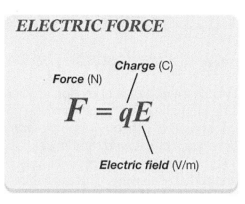

Challenge Questions!

b. The mass of an electron is 9.109 x 10⁻³¹ kg. Using your answer from the previous question, calculate the acceleration felt by the electron in this field. (*Hint*: Think about Newton's second law.)

c. Now that you know the acceleration that the electron is experiencing, how long will it take the electron to cross the 1 cm length of the field? (*Hint*: You have used the formula $d = 1/2at^2$ in other investigations. It can be used to solve for *t*, time.)

d. Now that you know how long it takes the electron to cross the field, and the acceleration it is experiencing, calculate its velocity when it goes through a hole in the screen. (*Hint*: Think about the definition of *acceleration*, and solve for velocity.)

6 Calculating force between charged particles

Protons and electrons are basic subatomic particles.

COULOMB'S LAW

$$F = K\frac{q_1 q_2}{R^2}$$

Force (N), Constant (9 x 10⁹ N·m²/C²), Charges (C), Distance (m)

Pennies are made of copper and zinc. A single penny is made of approximately 2 x 10²² atoms that contain 7 x 10²³ electrons and an equal number of protons.

a. A single proton or electron has a charge of 1.602 x 10⁻¹⁹ coulombs. How many coulombs of protons and electrons are in each penny?

b. Imagine you have two pennies, A and B. You remove all of the electrons from penny A and place them on penny B. Is penny A positively or negatively charged? Is penny B positively or negatively charged?

c. Calculate the force of attraction between the two pennies if they are held one meter apart.

d. A diesel locomotive weighs approximately one million newtons. Determine the number of locomotives necessary to have a weight equal to the strength of the force between the pennies.

You were probably surprised when you calculated the force between the pennies. What does your answer tell you about the amounts of charge normally transferred in activities such as brushing your socks on carpet or rubbing a balloon on your hair?

Investigation 18.1 Harmonic Motion and the Pendulum

18.1 Harmonic Motion and the Pendulum

How do we describe the back-and-forth motion of a pendulum?

Two common types of motion are linear motion and harmonic motion. One kind of motion goes from one place to another, like a person walking from home to school. This is *linear motion*. We use words such as *distance*, *time*, *speed*, and *acceleration* to describe linear motion. The second kind of motion repeats itself over and over, like a child going back and forth on a swing. This kind of motion is called *harmonic motion*. The word *harmonic* comes from the word *harmony,* meaning "multiples of." Any system that exhibits harmonic motion is called an *oscillator*.

In this investigation, you will:

- experiment with a pendulum and see what you can do to change the period.
- set up your pendulum to measure a 30-second time interval.

Materials List

From Springs and Swings:

- Small stand
- Mass hanger
- Washers
- Pendulum string
- Beam breaker

Other materials:

- DataCollector
- Photogate
- Measuring tape

1 A pendulum

A pendulum is an oscillator made from a mass on a string. The mass is free to swing back and forth.

- A *cycle* is one complete back-and-forth motion.
- The *period* is the time it takes to complete one full cycle. The period of a pendulum is the time it takes for the pendulum to swing from left to right *and back again.*
- The *amplitude* describes the size of the cycle. The amplitude of a pendulum is the amount the pendulum swings away from equilibrium.

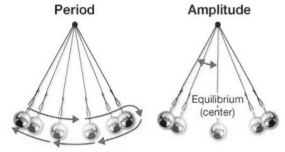

The period is the time to complete one cycle.

The amplitude is the maximum amount the system moves away from equilibrium.

2 Setting up the pendulum

1. Slide the pendulum string into the bracket at the top of the stand, as shown at right.
2. Attach the mass hanger to the loop end of the pendulum string. You now have a freely-swinging pendulum.
3. Attach the beam breaker to the string just below the bracket, as shown.
4. Attach a photogate to the top of the stand, as shown.
5. Plug the photogate into the A input on the DataCollector. Turn on the DataCollector and choose Timer mode from the home screen.

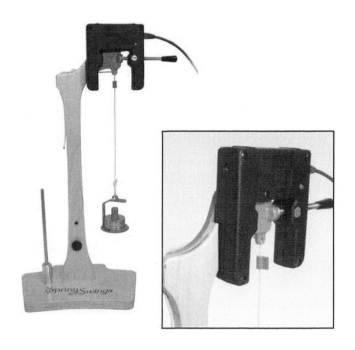

6. In Timer mode, choose the Period function (p). Period is the time for one cycle. The pendulum breaks the beam once as it swings through its complete cycle. The DataCollector's period readout corresponds directly to the time it takes the pendulum to complete one full cycle.

3 The three pendulum variables

In this experiment, the period of the pendulum is the dependent variable. There are three independent variables: the pendulum mass, the amplitude of the swing, and the length of the pendulum string.

1. Put a few washers on the mass hanger, adjust the string length to about 15 centimeters, and swing the pendulum through the photogate. Notice the period measured by the photogate.
2. The length of the string can be changed by removing it from the slotted bracket and placing it back in. You can change the mass by varying the number of washers on the mass hanger. The amplitude can be changed by varying the starting angle of the pendulum (low, medium, and high angle).
3. Design an experiment to determine which of the three variables has the greatest effect on the period of the pendulum. Your experiment should provide enough data to show that one of the three variables has a much greater effect than the other two. Be sure to use a technique that gives you consistent results. When you are changing a variable, it is a good idea to change it by a lot. For example, when you vary the mass, try experimenting with no washers, 5 washers, and 10 washers. You will only need to change each variable three times to see the effect. Be sure to design and use an appropriate data table.

4 Analyzing the data

a. Of the three things you can change (length, mass, and amplitude), which has the greatest effect on the pendulum, and why?

b. Split up your data by making three separate graphs so that you can look at the effect of each variable. To make comparison easier, make sure all the graphs have the same scale on the y-axis (period). The graphs should be labeled as shown in the example below:

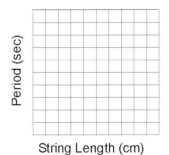
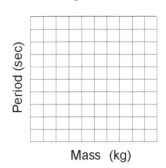
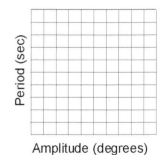

Investigation 18.1 Harmonic Motion and the Pendulum

5 Designing a solution

Pendulum clocks were once among the most common ways to keep time. You can still buy beautifully-made, contemporary pendulum clocks. For a pendulum clock to be accurate, its period must be set so that a certain number of periods equals a convenient measure of time. For example, you could design a clock with a pendulum that has a period of one second. The gears in the clock mechanism would then have to turn the second hand $1/60^{th}$ of a turn per swing of the pendulum.

a. Using your data, design and construct a pendulum that you can use to accurately measure a time interval of 30 seconds. Test your pendulum clock against the DataCollector stopwatch mode.

b. Mark on your graph the period you chose for your pendulum.

c. How many cycles did your pendulum complete in 30 seconds?

d. If mass does not affect the period, why is it important that the pendulum in a clock be heavy?

e. Calculate the percent error in your prediction of time from your pendulum clock.

18.2 Harmonic Motion Graphs

How do we make graphs of harmonic motion?

Harmonic motion graphs usually show cycles. The most common graph shows position versus time. Positive and negative positions are used to distinguish motion on either side of equilibrium. The amplitude is equal to one-half of the peak-to-peak distance on a graph. The period is read from the x-axis as the time for one cycle. It is useful to compare harmonic motion with circular motion. One rotation of a circle (360°) corresponds to one complete cycle. We often use degrees to indicate where a motion is in its cycle. For instance, 90° is one-fourth of a circle, so it represents one-fourth of a cycle. *Phase* describes where the motion is in a full cycle. Halfway through a cycle is a phase of 180°. Two harmonic motions are *in phase* when the phase difference between them is zero and out of phase when the phase difference between them is not zero. Motions that are 180° out of phase are always on opposite ends of their cycles.

In this investigation, you will:

- create graphs showing harmonic motion.
- use graphs of harmonic motion to determine period and amplitude.

1 Simple harmonic motion graph

Make a graph from the data table below. Draw a smooth curve that most closely shows the pattern in the data. Answer the questions about the graph.

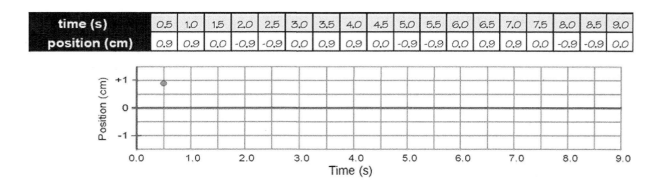

time (s)	0.5	1.0	1.5	2.0	2.5	3.0	3.5	4.0	4.5	5.0	5.5	6.0	6.5	7.0	7.5	8.0	8.5	9.0
position (cm)	0.9	0.9	0.0	-0.9	-0.9	0.0	0.9	0.9	0.0	-0.9	-0.9	0.0	0.9	0.9	0.0	-0.9	-0.9	0.0

a. What is the amplitude of the graph in centimeters?

b. What is the period of the graph in seconds?

Investigation 18.2 Harmonic Motion Graphs

2 Comparing harmonic motion graphs

Two different groups of students are doing experiments on giant pendulums with very long periods. Both groups use the same clock to record time. Below are two sets of position vs. time data, one from each group. Graph the data and answer the questions.

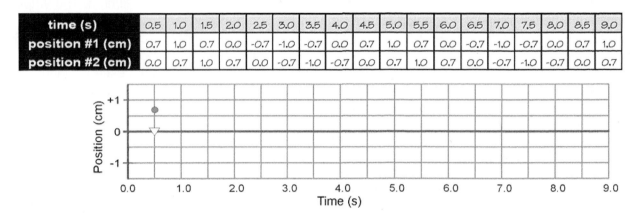

time (s)	0.5	1.0	1.5	2.0	2.5	3.0	3.5	4.0	4.5	5.0	5.5	6.0	6.5	7.0	7.5	8.0	8.5	9.0
position #1 (cm)	0.7	1.0	0.7	0.0	-0.7	-1.0	-0.7	0.0	0.7	1.0	0.7	0.0	-0.7	-1.0	-0.7	0.0	0.7	1.0
position #2 (cm)	0.0	0.7	1.0	0.7	0.0	-0.7	-1.0	-0.7	0.0	0.7	1.0	0.7	0.0	-0.7	-1.0	-0.7	0.0	0.7

a. Which pendulum was most probably released first? In your answer, you must use the word *phase* to explain how you chose which pendulum started first.

b. How much time was there between the start of the lead pendulum and the start of the other pendulum?

3 A complex harmonic motion graph

A clever inventor is trying to build a pendulum that can have two periods at the same time. The pendulum is actually two pendulums: a very light, short pendulum swinging from a much heavier long pendulum, as shown in the diagram below. The position vs. time graph for the combined pendulum was measured. The combined position is the position of the short pendulum while it is swinging from the long pendulum. The graph below shows the motion of the combined pendulum. Answer the three questions from this graph.

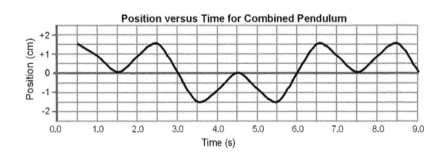
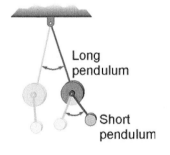

a. What is the amplitude of the combined pendulum in centimeters?

b. What is the period of the combined pendulum in seconds?

c. In one or two sentences, describe the difference between the graphs of the single pendulums and the graph of the combined pendulum.

Harmonic Motion Graphs   Investigation 18.2

4 How to get a complex graph

The inventor has a theory that the combined motion is the sum of the separate motions of the short and long pendulums. The inventor measured the position versus time for the short pendulum and the long pendulum separately. To test this theory, add up the positions for the short and long pendulums that were measured separately. For each point in time, write the sum on the line in the table below labeled "Long plus short." Graph the position versus time for the combined pendulum.

a. Does the graph confirm or disprove the inventor's theory? Explain your answer in a few sentences.

time (s)	0.5	1.0	1.5	2.0	2.5	3.0	3.5	4.0	4.5	5.0	5.5	6.0	6.5	7.0	7.5	8.0	8.5	9.0
Long Pendulum (cm)	0.5	0.9	1.0	0.9	0.5	0.0	-0.5	-0.9	-1.0	-0.9	-0.5	0.0	0.5	0.9	1.0	0.9	0.5	0.0
Short Pendulum (cm)	1.0	0.0	-1.0	0.0	1.0	0.0	-1.0	0.0	1.0	0.0	-1.0	0.0	1.0	0.0	-1.0	0.0	1.0	0.0
Long plus Short (cm)																		

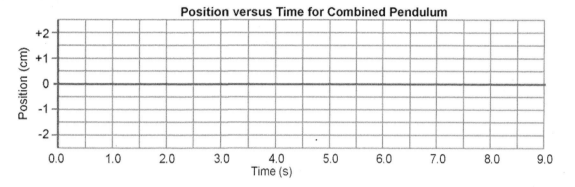

143

18.3 Natural Frequency

What kinds of systems oscillate?

Motion is the result when we disturb a system in equilibrium, and sometimes that motion is harmonic motion. For example, consider a cart balanced on top of a hill. If you push the cart even a little, it quickly rolls downhill and does not return. This is motion, but not harmonic motion. A cart in a valley is a good example of a system that *does* show harmonic motion. If you move the cart partly up the hill and release it, harmonic motion results as the cart rolls back and forth in the bottom of the valley.

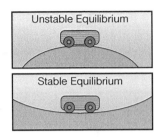

Materials List

From Springs and Swings:

- Small stand
- Mass hanger
- Beam breaker
- Washers
- Extension springs (blue tab and white tab)

Other materials:

- DataCollector
- Photogate

In this investigation, you will:

- build a mechanical oscillator.
- find the oscillator's period and natural frequency.
- change the oscillator's natural frequency.

1 Stable and unstable systems

The top of a hill is an example of *unstable* equilibrium. In unstable systems, forces act to pull the system away from equilibrium when disturbed. The bottom of a valley is an example of *stable* equilibrium. In a stable system, forces always act to restore the system to equilibrium when it is disturbed. We find harmonic motion in stable systems.

a. Describe an example of a stable system in one or two sentences. What happens when you push it a little away from equilibrium? Write one sentence that describes the motion.

b. Describe an example of an unstable system in one or two sentences. What happens when you push the unstable system a little away from equilibrium? Write one sentence that describes the motion.

2 Restoring forces

With a mechanical system like a pendulum, harmonic motion comes from the action of forces and inertia. A *restoring force* is any force that always tries to pull a system back toward equilibrium. For example, gravity always pulls the pendulum toward the center no matter which side it is on. Because it has inertia, the pendulum passes right through the middle and keeps going. The restoring force then slows it and accelerates the pendulum back toward equilibrium. However, the pendulum overshoots again and goes too far. The cycle repeats over and over to create harmonic motion (see below).

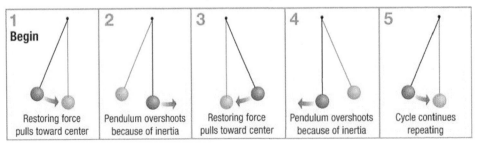

Natural Frequency Investigation 18.3

3 Building an oscillator

To make a mechanical oscillator, you need to provide some kind of restoring force connected to a mass that provides inertia. Rubber bands, strings, elastic bands, and springs can all provide restoring forces. Steel marbles, wood blocks, and even a rubber band have mass to supply inertia. The oscillation can go back and forth, up and down, twist, circle, or create any other motion that can be repetitive.

1. Set up a spring system that oscillates up and down (see photo at right). Use the extension spring that has the *blue* tab.
2. Pull the mass hanger down a very small amount and gently let go to create smooth oscillations.

a. Draw a sketch of your system and identify what makes the restoring force and where the mass is located that provides inertia.

4 The natural frequency

When you pluck a stretched string, it vibrates. If you pluck the same string 10 times in a row, it vibrates at the same frequency every time. The frequency at which objects tend to move in harmonic motion is called the *natural frequency*. Everything that can oscillate has a natural frequency.

1. Place 5 washers on the mass hanger, and place the photogate on the small stand as shown in the photo above.
2. Plug the photogate into the A input on the DataCollector. Select Timer mode, then the Period (p) function. For slow oscillators like this spring/mass system, rather than measuring frequency, it is more accurate to measure period and then *calculate* the natural frequency. Recall from the previous investigation that the period is the time it takes the oscillator to complete one cycle.

PERIOD AND FREQUENCY RELATIONSHIP

$$f_{\text{Frequency (Hz)}} = \frac{1}{T_{\text{Period (s)}}}$$

3. Gently pull down on the spring (you should be using the one with the blue tab) to begin a controlled, smooth oscillation. Practice until you can get smooth up-and-down oscillations with very little movement in any other direction. The period in seconds will vary slightly from trial to trial. Once you have good technique, record the period.

a. Calculate the natural frequency by dividing 1 by the period. Show your work.

Investigation 18.3 Natural Frequency

5 Changing the natural frequency

The natural frequency is a balance between the strength of the restoring force and the mass providing the inertia. To change the natural frequency, change the balance between force and inertia. You have another spring to try (it has a white tab) and you can vary the number of washers. You might have to place the photogate on the table as shown at right for some oscillator setups.

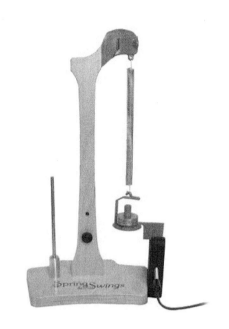

a. Describe and test a way to increase the natural frequency of your oscillator. Increasing the natural frequency makes the oscillator go faster.

b. Describe and test a way to decrease the natural frequency of your oscillator. Decreasing the natural frequency makes the oscillator go slower.

c. Create a data table to summarize your results.

d. Write a brief paragraph to explain *why* the changes you made to the oscillating system affected the natural frequency. Refer to evidence from your data table to support your explanation.

6 Natural frequency and guitar strings

a. Guitar strings are great examples of oscillators. To change the natural frequency of any oscillator, you must change the balance between force and inertia. A guitar player tightens and loosens individual string tensions to make each guitar string sound a certain way. When a string is tightened, its frequency increases, and the pitch of the string is higher. Explain why a guitar string's frequency increases when the tension increases, in terms of how the balance between force and inertia is changed.

b. If you look closely at an ordinary guitar, you will see that some strings are fatter and more massive than others. The more massive strings (at the bottom of the photo at right, showing a right-handed guitar) have a low pitch when you pluck them, and the smaller strings have a higher pitch. Explain why this is so in terms of changing the balance between force and inertia. You can refer to data you collected in Part 5 to support your answer.

19.1 Waves in Motion

How do waves move?

Waves are oscillations that move from one place to another. Like oscillations, waves also have the properties of frequency and amplitude. In this investigation, you will explore waves on strings and in water. What you learn applies to all other types of waves as well.

Materials List
- Metal toy spring
- Meter stick
- Food coloring
- Water
- Wave tray with aluminum blocks and plastic tube

1 Making a transverse wave pulse

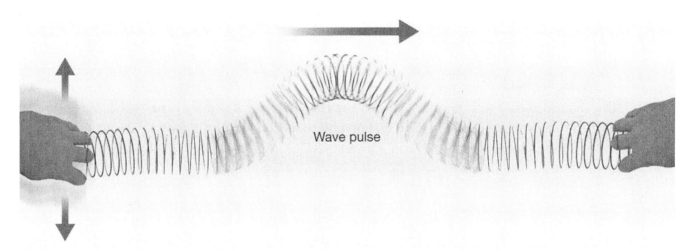

1. It takes two students to do this experiment. Each student takes one end of the spring.
2. Bring the spring down to the floor. Stretch it to a length of about three meters while keeping the spring on the floor.
3. One student should jerk one end of the spring rapidly to the side and back, just once. Make sure both ends of the spring are held tight and do not move once the wave is in motion. A wave pulse should travel up the spring.
4. Watch the wave pulse as it moves up and back. Try it a few times.

2 Constructing explanations

a. How is the motion of a wave pulse different from the motion of a moving object such as a car? (*Hint*: What is it that moves, in the case of a wave?)

b. What happens to the wave pulse when it hits the far end of the spring? Watch carefully. Does the pulse stay on the same side of the spring or flip to the other side? Use the word *reflect* in your answer.

c. Imagine you broke the spring in the middle. Do you think the wave could cross the break? Discuss the reasoning behind your answer in a few sentences.

d. Why does the wave pulse move along the spring instead of just staying in the place where you made it?

Investigation 19.1 Waves in Motion

3 Making a longitudinal wave pulse

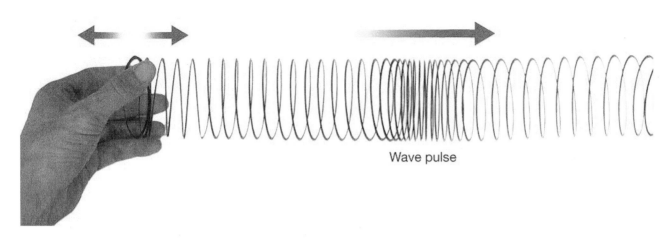

Wave pulse

1. Just like in Part 1, each student should take one end of the spring.
2. Bring the spring down to the floor. Stretch it to a length of about three meters while keeping the spring on the floor.
3. One student should jerk one end of the spring rapidly forward and back, just once. Make sure both ends of the spring are held tight and do not move once the wave is in motion. A wave pulse should travel up the spring.
4. Watch the wave pulse as it moves up and back. Try it a few times.

4 Constructing explanations

a. How is the motion of the longitudinal wave pulse different from the transverse wave pulse you made in Part 1? (*Hint*: How is the motion of the spring itself different?)

b. What happens to the wave pulse when it hits the far end of the spring? Does it behave like the transverse wave, or differently? Use the word *reflect* in your answer.

c. Why do you think longitudinal waves are also sometimes called "compression waves"?

d. Do you think a wave can be made only by stretching the spring instead of compressing it? Make a prediction, and then try it to see if you were right.

Waves in Motion Investigation 19.1

5 Waves in water

1. Fill a flat tray with about one-half centimeter of colored water. The color helps you see the waves.
2. Roll the wave tube forward about 1 centimeter in a smooth motion. This launches a nearly straight wave called a *plane wave* across the tray.
3. Next, poke the surface of the water with your fingertip. Disturbing a single point on the surface of the water makes a *circular wave* that moves outward from where you touched the water.
4. Arrange two blocks so they cross the tray leaving a 1-centimeter opening between them.
5. Make a plane wave that moves toward the blocks. Observe what happens to the wave that goes through the opening.

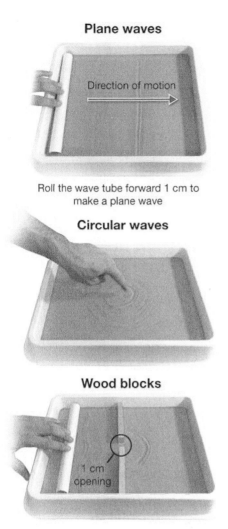

6 Constructing explanations

a. Draw a sketch that shows your plane wave from the top. Also on your sketch, draw an arrow that shows the direction in which the wave moves.
b. Is the wave parallel or perpendicular to the direction the wave moves?
c. Draw another sketch that shows the circular wave. Add at least four arrows that show the direction in which each part of the wave moves.
d. At every point along the wave, are the waves more parallel or perpendicular to the direction in which the circular wave moves?
e. Sketch the shape of the wave before and after passing through the one-centimeter opening.
f. Does the wave change shape when it passes through the opening? Describe what you observed.
g. Are the waves you made in the water transverse or longitudinal waves? How do you know?

149

19.2 Resonance and Standing Waves

How do we make and control waves?

The vibrating string is perfect for investigating waves because the waves are large enough to see easily. What you will see and learn applies to guitars, pianos, drums—almost all musical instruments.

In this investigation, you will:

- explore the relationship between the frequency of a wave and its wavelength.

Materials List
- Sound and Waves kit
- Physics Stand
- DataCollector

1 Setting up the experiment

Connect the DataCollector to the wave generator as shown in the diagram. The telephone cord connects the DataCollector and wave generator. The black wire goes between the wave generator and the wiggler.

1. Attach the fiddlehead to the top of the stand as high as it will go.
2. Attach the wiggler to the bottom of the stand as low as it will go.
3. Stretch the elastic string a little (5 to 10 centimeters) and attach the free end to the fiddlehead. Loosen the knob until you can slide the string between any two of the washers. *Gently* tighten the knob just enough to hold the string.
4. Turn on the DataCollector and be sure to plug in the AC adapter.
5. Set the wave generator to Waves using the button. The wiggler should start to wiggle back and forth, shaking the string.
6. Set the DataCollector to measure frequency. You should get a reading of about 10 Hz, which means the wiggler is oscillating back and forth 10 times per second.
7. Try adjusting the frequency of the wiggler with the frequency control on the wave generator. If you watch the string, you will find that interesting patterns form at certain frequencies.

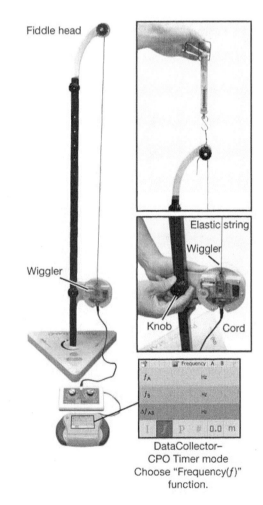

DataCollector–
CPO Timer mode
Choose "Frequency(*f*)" function.

2 Waves on the vibrating string

The *frequency* (*f*) is the rate at which the string shakes back and forth, or oscillates. A frequency of 10 Hz means the string oscillates 10 times each second.

At certain frequencies, the vibrating string forms wave patterns called *harmonics*. The first harmonic has one bump, the second harmonic has two bumps, and so on.

The *wavelength* (λ) is the length of one complete wave. One complete wave is two "bumps." Therefore, wavelength is the length of two bumps. The string is one meter long. If you have a pattern of three bumps, the wavelength is two-thirds of a meter, since three bumps equal one meter and a wave is two bumps.

The *amplitude* (A) of the wave is the maximum amount the string moves away from its resting (center) position. A *node* is a point where the string does not move. An *antinode* is a point where the amplitude is greatest. You can measure the wavelength as the distance separating three consecutive nodes.

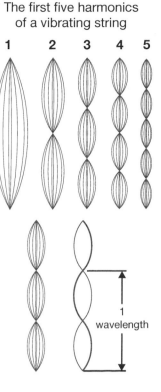

The first five harmonics of a vibrating string

1 wavelength

Create a table like Table 1 (below) in your notebook. Adjust the frequency to obtain the first eight to ten harmonics of the string and record the frequency and wavelength for each one in your Table 1. You should fine-tune the frequency to obtain the largest amplitude before recording the data for each harmonic. Look for harmonics 2 to 6 before looking for the first one. The first harmonic, also called the *fundamental*, is hard to find with exactness. Once you have the frequencies for the others, they provide a clue for finding the frequency of the first harmonic.

Table 1: Frequency and wavelength of first six harmonics

Harmonic #	Frequency (Hz)	Wavelength (m)	Frequency times wavelength
1			
2			
3			

3 Arguing from evidence

a. In one or two sentences, describe how the frequencies of the different harmonic patterns are related to each other.

b. Why is the word *fundamental* chosen as another name for the first harmonic?

c. Give an equation relating frequency (*f*) and wavelength (λ) that best describes your observations.

d. If the frequency increases by a factor of two, what happens to the wavelength?

e. Propose a meaning for the number you get by multiplying frequency and wavelength.

19.3 Exploring Standing Wave Properties

How does changing the tension affect a vibrating string?

A guitar uses vibrating strings to produce music. The guitar player tightens and loosens individual string tensions to make each guitar string sound a certain way. What happens to the frequency of the vibrating guitar string when its tension is increased? Why do guitar strings have different thicknesses and masses? What happens to the amplitude of a standing wave when you change the tension? This investigation will provide the answers.

Materials List
- Sound and Waves kit
- Physics Stand
- DataCollector
- Spring scale

In this investigation, you will:

- explore the relationship between the tension of a vibrating string and its frequency and amplitude.

1 Doing the investigation

Connect the DataCollector to the wave generator as shown.

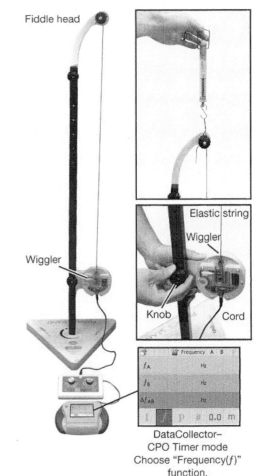

1. Place the physics stand on the floor. Attach the fiddlehead to the top of the stand as high as it will go.
2. Attach the wiggler to the bottom of the stand as low as it will go.
3. Secure the knot-end of the elastic string so it is caught in the wiggler arm notch.
4. Loosen the knob on the fiddlehead. Slide the free end of the string through the washers. It should move freely because the knob is still loose. Stretch the string up beyond the fiddlehead a little.
5. Attach a spring scale to the free end of the elastic string.
6. While the knob is loose, stretch the string until the spring scale reads 0.5 N.
7. Gently tighten the knob to set the string without changing the tension.
8. Adjust the frequency of the wave generator until you have the third harmonic wave. Remember to fine-tune the frequency until the amplitude of the wave is as big as it can get.
9. Record the frequency in Table 1.
10. Repeat steps 4 through 9 for each different string tension.
11. Observe what happens to the wave amplitude in each trial.

Table 1: Changing the string's tension

Harmonic #	Tension (N)	Frequency (Hz)
3	0.5	
3	1.0	
3	1.5	
3	2.0	
3	2.5	
3	3.0	

2 Constructing explanations

String tension versus frequency

a. What happens to the natural frequency as you increase the tension of the string? In your answer, discuss why this is useful in tuning a stringed instrument such as a guitar or piano.

b. In a previous investigation, you learned that a balance between inertia and restoring force is what determines the natural frequency of an oscillator like a vibrating string. Explain *why* the frequency changed the way it did when you tightened the string, in terms of how the balance between force and inertia was changed.

String mass versus frequency

c. Suppose you used a much thicker, more massive elastic string for this experiment. How would the results of the experiment using this string compare to the data in Table 1? Explain.

d. A basic guitar has six strings, and each is a different thickness. Based on your answer to question c, how does the pitch of the sound from a more massive string compare to the pitch of the sound from the thinnest guitar string?

e. Which of Newton's three laws of motion best explains your answers to questions c and d? Discuss.

String tension versus amplitude

f. As you increased the tension, making the string stiffer, what happened to the amplitude of the wave?

g. An earthquake is like the wiggler in that it makes the ground shake back and forth with a certain frequency. How do your results relate to making tall buildings sway less in an earthquake? You should consider what happened to the amplitude of the wave when you increased the tension in the string to answer the question.

h. Steel and masonry are two common building materials used around the world. Use the Internet to look up the term *ductility* and how it applies to the use of both steel and masonry in building construction. Why would one or the other be better, or safer, to use in earthquake-prone areas around the world?

20.1 Sound and Hearing

What is sound and how do we hear it?

The ear is a remarkable sensor. Sound waves that we can hear change the air pressure by only one part in a million.

In this investigation, you will:

- learn about the range of frequencies the ear can detect and how small a difference in frequency we can perceive.
- use techniques for making experiments on human perception reliable.

Materials List
- Sound generator
- Speakers
- DataCollector

1 How high can you hear?

The accepted range of frequencies the human ear can hear ranges from a low of 20 Hz to a high of 20,000 Hz. There is tremendous variation within this range, as people's hearing changes greatly with age and exposure to loud noises.

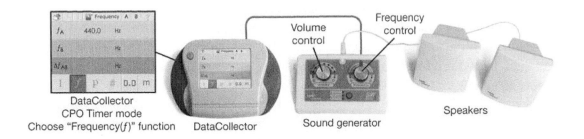

1. Connect the speakers to the sound generator. Connect your sound generator to a DataCollector. Choose Timer mode and select the Frequency function. You should hear a sound and the DataCollector should display a frequency near 440 Hz.
2. The sound generator has dials for frequency and volume control. Try adjusting the frequency and see how high and low it goes.

See if you and your group can agree on frequencies at which you hear the sound as low, medium, high, or very high frequency. Use Table 1 to record these frequencies. Don't try to be too exact, because the words *low*, *medium*, and *high* are not well defined. It is difficult to agree exactly on anything that is based completely on individual human perception.

Table 1: Perceived sound frequencies

Description	Frequency (Hz)
Low	
Medium	
High	
Very high	

2. Testing the upper frequency limit of the ear

Your teacher will use a sound generator to make sounds with frequencies up to 20,000 Hz. When the teacher asks, raise your hand if you can hear the sound. Do not raise your hand if you cannot hear it. Someone will be appointed to count hands and survey the class to see what fraction of students can still hear the sound.

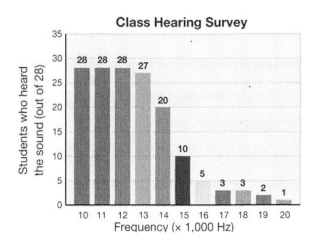

a. Make a histogram showing your class response to frequencies between 10,000 and 20,000 Hz. You should have 10 bars, one per 1,000 Hz. Each student who raises a hand is counted as a positive response on the graph.

b. Do you think the method of counting raised hands is likely to give an accurate result? Give at least one reason you believe the method is either good or bad.

20.2 Properties of Sound Waves

Does sound behave like other waves?

What experimental evidence is there that sound has the properties of waves? The fact that sound can produce interference with a characteristic wavelength is strong evidence that sound is a wave.

In this investigation, you will:

- experiment with a sound wave interference phenomenon called "beats."
- use interference to measure the wavelength of sound.

Materials List
- Sound generator
- Speakers
- DataCollector
- Meter stick
- Tuning fork

1 Beats

When you listen to a pair of speakers, two sounds reach your ear at the same time. What you hear is the combination of the two waves.

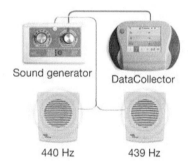

1. Set up the sound generator, speakers, and DataCollector as shown at right. Choose Timer mode and select the Frequency function.
2. When the sound generator is in Sound mode, you will hear a base frequency of 440 Hz coming from both speakers. Switch the unit from Sound mode to Beats mode by pressing the Mode button.
3. In Beats mode, turning the frequency dial changes the frequency of the sound produced by the second speaker from 440 Hz to 439 Hz and as far down as 430 Hz. The first speaker continues to play 440 Hz. In Beats mode adjust the frequency to read 439 Hz.
4. Stand back and listen to the sound when both are at equal volume.
5. Cover one speaker with your hand and turn it away so you can listen to the uncovered speaker by itself. Swap speakers and do the same so you can listen to the other speaker by itself. Can you tell the difference between the sound produced by each speaker?
6. Listen to both at equal volume again and listen to the combination.
7. Adjust the frequency between 439 and 430 Hz. Listen to the combinations of sound.

When two waves are close, but not exactly matched, in frequency, you hear beats. Beats are an example of the interference of sound waves. The sound gets alternately loud and soft as the waves drift in and out of phase with each other. Sometimes they are in phase and the result is twice as loud. A moment later, they are out of phase and they cancel out, leaving periods of quiet. The alternation of loud and soft is what we hear as beats.

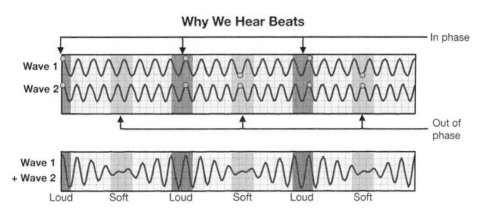

2 Interference

Suppose you have two identical sound waves and you are standing where you can hear them both. For certain positions, one sound wave reaches your ear in the opposite phase from the other wave, and the sound gets softer. Move over a little and the two sound waves add up to get louder. These effects are another example of interference and are easy to demonstrate.

1. Put the sound generator back in Sound mode. Place one speaker about 1/2 meter behind the other.
2. Set the frequency between 400 and 800 Hz.
3. Stand 3 or 4 meters in front of one speaker and have your lab partner slowly move the rear speaker away from you. You will hear the sound get loud and soft and loud again when the distance between speakers has changed by one wavelength.

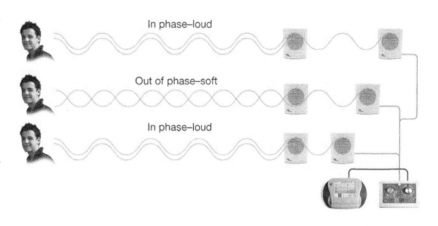

When two speakers are connected to the same sound generator in Sound mode, they both make the exact same sound wave. If you move around a room, you will hear places of loud and soft volume based on the difference in phase of the sound waves produced by the two speakers.

a. Try to make an approximate measure of the wavelength of sound by changing the separation of the two speakers. The speakers have been moved one wavelength when the sound heard by the observer has gone from loudest, to softest, and back to loudest again. For this to work, you need to keep the observer and both speakers in the same line.

b. Interference can be bad news for concert halls. People do not want their sound to be canceled out after they have bought tickets to a concert! Why do we usually not hear interference from sound systems even when they only have two speakers?

Investigation 20.3 Sound as a Wave

20.3 Sound as a Wave

How can we observe sound as a wave? How can we use the speed of sound and certain frequencies to build a basic instrument based on wavelength?

Waves are considered a traveling form of energy because they can cause changes in the objects they encounter. Sound waves carry energy that causes human eardrums to vibrate, and the brain interprets these vibrations as sound. Because it is a wave, sound has both frequency and wavelength. In this investigation, you will use a sound's frequency and its speed to calculate desired wavelengths and use your calculations to build a musical instrument.

Materials List
- 10-foot length 1/2-inch PVC pipe
- Metric ruler
- Pipe cutter (or a saw)
- Sandpaper
- Permanent marker
- DataCollector
- Temperature probe

1 Thinking about sound as a wave

You have learned several important properties of waves. Answer the questions below to review what you have learned.

a. What property of waves describes the distance from any point on a wave to the same point on the next cycle of the wave?

b. What property of waves describes how often something repeats, expressed in hertz?

c. How are the frequency and wavelength of a wave related to its speed?

d. How is the wavelength of a sound wave related to the pitch it produces?

2 Making a prediction

You have been given a long length of pipe. In Part 3, you will cut two lengths of pipe, one twice as long as the other. You will then use the pipe to produce a sound by blowing over the top of the pipe with your hand covering the bottom, or by tapping the bottom of the pipe on the palm of your hand. Both methods produce a sound with the same pitch. The air inside the pipe vibrates to make the sound you hear when the pipe is played. You may have to practice to get a repeatable sound.

a. Make a prediction about how the sound produced by the two different lengths of pipe will compare. Describe the sound in terms of pitch.

b. Explain the reasoning behind your prediction.

3 Cutting two sections of pipe

1. A good rule to follow when cutting anything is "measure twice, cut once." From the end of the pipe, measure a distance of 10.5 centimeters and use a pen or pencil to mark the distance with a small straight line. Depending on the cutting tool you have, follow your teacher's directions for how to cut the pipe as close to 10.5 centimeters as possible.

2. Once it is cut, use some sandpaper to remove any rough edges from the cut end of the pipe. It does not need to be perfectly smooth, but there should be no sharp edges.

3. When your first pipe is cut and sanded, measure 21 centimeters from the cut end of the pipe and make a straight line mark. Double-check your measurement and once again follow your teacher's directions on how to cut the pipe.
4. Sand both ends of the 21-centimeter pipe.

4 Using sound as an observation

You are ready to use your pipes to make sounds. Blow into both pipes and listen to the pitch each one makes. After you have compared the sounds created by each, answer the questions below.

a. Which pipe had a higher pitch, the short pipe or the long pipe?
b. Was your prediction correct about the pitch produced by two different lengths of pipe?
c. How does the pitch relate to the frequency of the sound created by both pipes?
d. How does the pitch relate to the wavelength of the sound created by both pipes?
e. How would the pitch of a pipe longer than 10.5 centimeters and shorter than 21 centimeters compare to the pitches of these two pipes?
f. If you were to cut increasingly longer pipes, how would the sound produced by each longer pipe be different from the previous shorter length? Describe the change in terms of frequency and wavelength.

5 Wavelengths based on the speed of sound and frequency

The two pieces of pipe you cut have different pitches. Each different pitch is called a *note*. Certain notes sound pleasant to the human ear, and these can be arranged together to create a musical scale. By using what you know about the relationship between the speed of sound, frequency, and wavelength, calculate the wavelength of the sound wave that will make each note in a series of musical scales. The first task is to calculate the speed of sound in your classroom. The speed of sound depends mainly on the temperature of the air in your classroom. Use the formula to calculate the speed of sound in air for your classroom based on the temperature in degrees Celsius.

$$\text{Speed of sound} = 331 \text{ m/s} + (0.6 \text{ m/s/°C} \times \text{°C})$$

Speed of sound m/s	=	331 m/s	+(0.6 m/s/°C	x	°C in classroom)
	=	331 m/s	+(0.6m/s/°C	x	°C)

Each note in all musical scales corresponds to a particular frequency. These notes and their frequencies have been established for thousands of years. Once you have calculated the speed of sound in your classroom, you can begin to calculate the wavelength of the sound waves needed to create the frequencies that make the musical scales we will use. Fill in your speed of sound on Table 1, and use the frequencies provided to calculate the wavelengths.

Investigation 20.3   Sound as a Wave

Table 1: Calculating wavelengths needed based on speed of sound and frequency

Speed of sound (m/s)	÷	Frequency (Hz)	=	Wavelength (m)
	÷	349	=	
	÷	392	=	
	÷	440	=	
	÷	466	=	
	÷	523	=	
	÷	587	=	
	÷	659	=	
	÷	698	=	
	÷	784	=	
	÷	880	=	
	÷	892	=	
	÷	1049	=	
	÷	1174	=	
	÷	1318	=	
	÷	1397	=	

6 Creating sound from a pipe with one closed end and one open end

The two pipes you've made have one closed end and one open end when they make sound. The air inside the pipe vibrates to make the note you hear when the pipe is played. Air in a tube like this vibrates in a specific way. The vibrations of the sound wave resonate in the pipe, amplifying their perceived volume. The length of the sound wave created by each pipe is actually four times the length of the pipe, as shown in the diagram below.

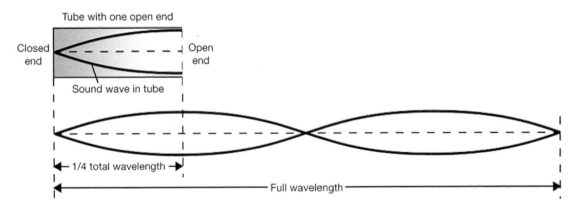

a. How would you calculate the length of pipe needed that would create a note of a specific frequency based on your calculated wavelengths from Part 5?

b. If the temperature in your room went up by 10°C, how would that change the length of pipe you need to cut to make a note of 440 Hz, compared to the length you would need for the current temperature of your classroom?

7 Cutting the remaining pipe

In Table 2, fill in the wavelengths of sound you calculated in Table 1 to make the frequencies needed at your classroom's temperature. The frequencies are now also labeled with the notes they make in the musical scale. Use your answer from question 6a above to calculate the length of pipe needed to make each note in the scale. Once you have calculated the length of pipe needed in meters, convert your calculation to centimeters and fill in the last column of Table 2. Your teacher will assign some notes to your group, and you will cut those notes' specific lengths of pipe. Use the pipe measurements in centimeters to measure and cut the lengths of pipe assigned to your group. You may be able to use the two pipes you made earlier. They may need to be trimmed to fit the particular lengths you need. Follow your teacher's instructions to cut the pipe. Use a permanent marker to label each pipe with the corresponding note it creates.

Table 2: Calculating the length of pipe for the required frequencies

Note in scale	Frequency (Hz)	Wavelength of note (m)	Length of pipe (m)	Length of pipe (cm)
F	349			
G	392			
A	440			
BFlat	466			
C	523			
D	587			
E	659			
F	698			
G	784			
A	880			
BFlat	892			
C	1,049			
D	1,174			
E	1,318			
F	1,397			

8 Playing some music

Your teacher will have some scripted music that you can play. Look through the song to see what notes will be played. That way you can collect all the pipes you will need during the song and have them ready to go. With members each playing one or two pipes, your group should be able to cover all the notes in a song. Melody and harmony are different parts of the song that work together to make the song more interesting to listen to. Get your pipes ready and follow the beat of the conductor. Good luck!

9 Using your model

a. There are some notes on the scale that have the same name but different frequencies. What is the relationship between these notes in terms of their frequencies and their wavelengths?

b. What do we call the group of eight notes that are arranged in increasing or decreasing pitch, starting with one note and ending with the note of the same name but different frequency? (For example, from F 349 to F 698.)

c. What is the difference between the vibration of air in a tube with one open end and the vibration of a plucked string?

d. **Design a solution:** Use the extra pieces of pipe that are left over to create a musical instrument capable of producing several different notes. How your instrument changes between notes is up to you. Experiment with different ideas until you come up with something. You may need to include other parts your teacher may provide for you to use. Once you have an instrument that works, think about how you could make it better. Would having more pipe help you? Would being able to cut fresh pieces from a long, uncut piece help you? Are the different notes your instrument produces actually part of a musical scale? Are the notes all part of the same musical scale? How can you test them? How can you tune your instrument? Try and see if you can make your instrument better by employing the engineering cycle of design, prototype, test, evaluate, and repeat.

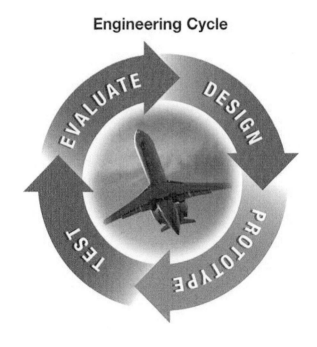

21.1 Properties of Light

What are some useful properties of light?

It is hard to imagine living in a world without light. Light has many properties that make it important for all living things.

In this investigation, you will:

- observe interesting examples of the reflection and refraction of light.
- compare the speed of light with the speed of sound.
- calculate the approximate intensity of the Sun's light.

Materials List

From Optics kit:
- Laminated sheet
- Red filter cap
- Mirror
- Prism

Miscellaneous:
- Index card
- Small piece of paper
- Marker
- Calculator
- Ruler

1 Fooling the brain with light

Prisms, mirrors, and lenses are devices that can affect the way light travels from an object to your eyes. Your brain perceives an object to be located wherever the light from that object *appears* to come from. The object may not actually be in that location, but your brain tells you it is there.

1. Set one of the filter caps on the laminated graph sheet. Fold a large index card in half lengthwise and set it on the graph sheet so that it blocks your view of the cap.

2. Arrange the mirror so that you can see the cap on the other side of the folded paper.

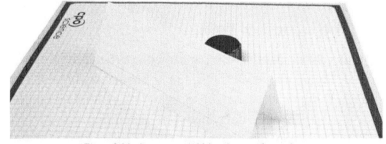
Place folded paper so it hides the cap from view.

Now arrange the mirror so you can see the cap, even though it is hidden from view behind the paper.

a. Draw a sketch that uses lines to show how the light from the cap reaches your eyes by reflecting from the mirror.

b. Where does the cap appear to be when you see its reflection in the mirror: at the surface of the mirror, in front of the mirror, or behind it? Give evidence for your answer.

163

Investigation 21.1 Properties of Light

2 Seeing reflection and refraction at the same time

Both reflection and refraction often occur when light hits a boundary between materials, such as the boundary between glass and air. The amount of light reflected or refracted depends on the angle at which you are looking relative to the surface.

Fold a paper card marked with "A" and "B"

The image in the prism changes as you move your head.

1. Take a piece of paper or index card about the size of a business card and fold it in half lengthwise. Draw the letter *A* on one side of the fold and the letter *B* on the other side (see photo above).
2. Wrap the folded paper around one of the corners of the prism that is *not* a right angle.
3. Look into the prism. Move your head up and down to change the angle at which you look.

a. Draw a diagram showing the path of the light when you see the letter *A*.

b. Draw a diagram showing the path of the light when you see the letter *B*.

c. Is the image in the prism always reflected or refracted, or can there be both reflection and refraction at the same time? Give evidence from the prism experiment to support your answer.

3 The speed of light

Light travels so fast that it is hard to comprehend the tiny interval of time it takes to get from one place to another. Sound also travels fast, but not as fast as light. You can perceive effects due to the difference between the speeds of sound and light. For example, you see a distant lightning bolt before you hear the thunder clap. The time delay between seeing lightning and hearing its sound (the thunder) is due to the time it takes sound to travel to your ears. The speed of light is so fast (3×10^8 m/s) that there is no apparent delay between the flash of a distant lightning bolt and the moment the light rays reach your eyes.

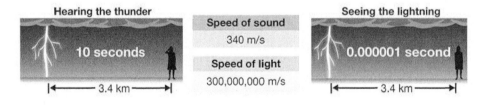

Distances between locations on Earth are very small compared to distances between Earth and stars. Light travels so fast that we don't notice the time light takes to travel between two places on Earth. But what about the light from farther away? How long does it take light to reach your eye from the Sun? As we look out into the universe, we are also seeing backward in time because the light from distant stars and galaxies has traveled many years before reaching human telescopes. Using this technique, we are able to see more than 1 billion years back in time.

a. Find out how many kilometers Earth is from the Sun, on average.

b. Calculate the time it takes light to travel this distance.

c. Calculate how far light travels in one second.

21.2 Additive Color Model and Vision

How do we see color?

All the colors of visible light can be created using a combination of three primary colors: red, blue, and green. You will use a white light source and color filters to discover what happens when you mix different colors of light. You will also learn how those filters work.

In this investigation, you will:

- show that white light can be made from red, green, and blue light.
- explain the colors we see in terms of subtracting colors from white light.

Materials List

From Optics kit:
- Red, green, and blue flashlights with holders
- Light blue lens
- Diffraction glasses (optional)

1 Mixing primary colors of light

1. Slide all three flashlights into their own holders.
2. Turn the flashlights on.
3. Connect the red and green flashlights by sliding their holders together using the rail and slot connectors on the side.

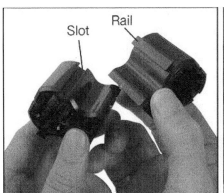
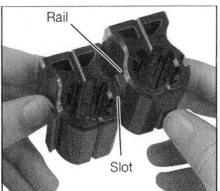

Investigation 21.2 Additive Color Model and Vision

4. Place the blue flashlight on top of the red and green lights, making a small pyramid stack. Set the blue light on top of the other two with the holder on its side, so the rail on the holder fits in the small groove created between the holders of the red and green lights.

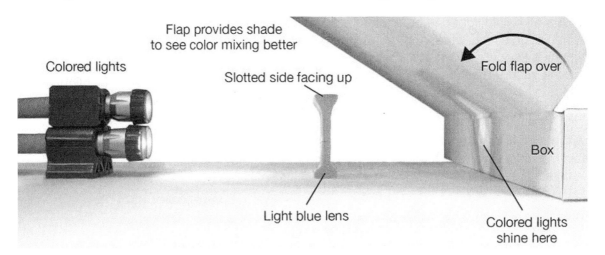

5. Place the light blue lens, slotted side facing up, just in front of the lights so they shine through it. This will increase the visibility of the color mixing.
6. Set the white box that contained the Optics with Light & Color kit on the opposite side of the lens from the lights. Fold the top of the box over to shade the area where the three colored lights are shining on the side of the box.
7. Slowly move the lens away from the lights and toward the box until you see the three spots of color (red-green-blue) overlap on the box.

2 Constructing explanations

Use Table 1 to record the answers to the following questions about your observations.

a. What color do you see when you mix red and green light?
b. What color do you see when you mix green and blue light?
c. What color do you see when you mix blue and red light?
d. What color is produced when all three colors of light are equally mixed?

Table 1: Mixing primary colors of light

LED color combination	Color you see
Red + green	
Green + blue	
Blue + red	
Red + green + blue	

3 Sources of light

a. Televisions, computer monitors, and cell phone display screens use light to create images. What colors of light can these devices use to make all the other possible colors needed for the images they display?

b. How do you think they use these colors to create images?

c. Look at your clothes. Does the light reaching your eyes from your clothes originate in your clothes, or does it come from somewhere else?

d. What color would your clothes appear to be if the room you were in was completely dark?

e. What is the difference between the light that reaches our eyes from a colorful painting, and an image you see of that same painting displayed on a computer screen?

4 Seeing Colors

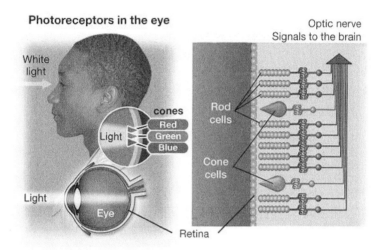

a. Research and explain the following terms from the diagram above: *cone cells*, *rod cells*, *retina*, and *optic nerve*.

b. Research and explain how the eye sees green light in terms of the photoreceptors in the eye.

c. Research and explain how the eye sees yellow light in terms of the photoreceptors in the eye.

d. Research and explain how the eye sees white light in terms of the photoreceptors in the eye.

21.3 Subtractive Color Model

How is color created by light?

White light is made up of all the colors of the visible spectrum. Most places, such as your home, office buildings, and your classroom, are illuminated by white light. This investigation will examine how the color an object appears to be depends on the colors of light it absorbs and reflects.

In this investigation, you will:

- mix different pigments and observe how subtractive color mixing works.
- explain the colors we see in terms of subtracting colors from white light.

Materials List
- Color Mixing set

1 The subtractive color model (CMYK)

1. You have three colors of clay: yellow, magenta, and cyan. Take a portion the size of your fingertip of the both cyan and the magenta. Mix them together. What color do you get?
2. Mix equal amounts of cyan and yellow. What color do you get?
3. Mix equal amounts of yellow and magenta. What color do you get?

The subtractive color model (CMYK)

	Cyan	Magenta	Yellow	Black
Absorbs	Red	Green	Blue	Red, Green, Blue
Reflects	Blue, green	Blue, red	Red, Green	None

Mix equal amounts of the three subtractive primary colors (two colors at a time)

2 Constructing explanations

a. Explain how the mixture of magenta and cyan makes its color when seen in white light.
b. Explain how the mixture of cyan and yellow makes its color when seen in white light.
c. Explain how the mixture of yellow and magenta makes its color when seen in white light.
d. Why don't the mixed colors produce full red, green, or blue?
e. How is the additive color model different from the subtractive color model?
f. Research how printers make colors. Do they use red, green, and blue (RGB) or cyan, magenta, yellow, and black (CMYK)? Explain why printed pictures need to use one or the other.
g. Research how computer monitors and televisions make colors. Do they use red, green, and blue (RGB) or cyan, magenta, yellow, and black (CMYK)? Explain why TVs and computer screens need to use one or the other.

22.1 Reflection

How do we describe the reflection of light?

We observe the law of reflection every day. Our sense of sight depends on light reflected from objects around us.

In this investigation, you will:

- take a closer look at reflection.
- apply geometry to predict exactly where reflected light goes.
- determine the rules for how and to what degree light is reflected.

Materials List

From Optics kit:
- Laminated graph sheet
- Laser flashlight and holder
- Mirror

Miscellaneous:
- Water-soluble marker
- Protractor

 Avoid shining a laser directly into the eye.

1 Observing the law of reflection

How to trace the beam of the laser

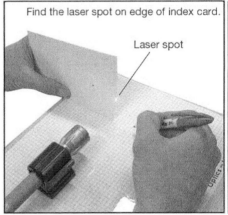
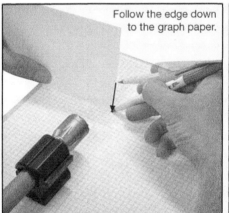

1. Place your laminated graph sheet on a flat surface and align the laser so the beam follows one horizontal line across the paper.
2. Set the mirror on the laminated sheet so the light beam from the laser hits its shiny surface at an angle. Draw a line on the graph paper, marking the position of the front face of the mirror.
3. Use a fine-point, water-soluble marker and an index card to trace the incident and reflected light rays from the laser. See the photos above and at right for clarification.
4. Repeat steps 1–3 with the mirror set at a different angle. Do the experiment for at least four different angles. Use different colored markers or label your incident and reflected rays so you don't get them confused.

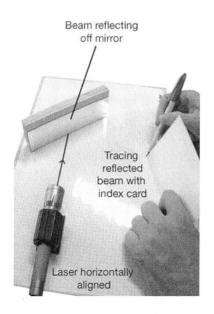

Investigation 22.1 Reflection

2 Constructing explanations

A diagram showing how light rays travel is called a *ray diagram*. Lines and arrows on a ray diagram represent rays of light. The *incident* ray travels to the mirror, the *reflected* ray travels away from the mirror.

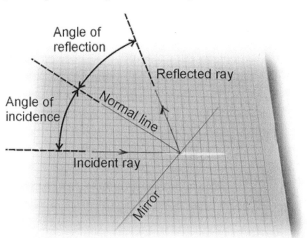

a. For each ray diagram, draw a line perpendicular to the mirror surface at the point where the rays hit. This line is called the *normal line*.

b. Use a protractor to measure the angle between the normal and the incident and reflected rays. Record your measurements in Table 1.

c. Write down your own statement of the law of reflection, describing the relationship between the angles you measured.

d. A laser shines at a mirror at an angle of incidence of 75 degrees. Predict its angle of reflection. After predicting, test your prediction. Were you right?

Table 1: Angles of incidence and reflection

	Diagram #1	Diagram #2	Diagram #3	Diagram #4
Angle of incidence				
Angle of reflection				

22.2 Refraction

How do we describe the refraction of light?

Light rays can bend when they cross an interface between two different materials, like air and glass, or air and water. The bending of light rays at a boundary between materials is called *refraction*. Prisms and lenses use refraction to manipulate light in telescopes, binoculars, cameras, and even your eyes.

In this investigation, you will:

- take a closer look at refraction.
- trace light rays before and after refraction and examine how they change direction.
- experiment with the critical angle of refraction.

Materials List

From Optics kit:
- Laminated graph sheet
- Laser flashlight and holder
- Prism

Miscellaneous:
- Water-soluble marker
- Protractor
- Graph paper and pencil
- Clear plastic cup
- Water
- Drop of milk

  Avoid shining a laser directly into the eye.

1 Refraction

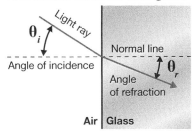
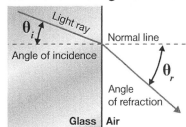

The normal line is used to describe how refraction works. The normal line is a line perpendicular to the surface. A light ray falling on a surface is called the *incident ray*. The light ray passing through the surface is called the *refracted ray*. Light rays move in straight lines except at the point where they cross the surface. Find the incident and refracted rays on the diagram above.

The *angle of incidence* is the angle between the incident ray and the normal line. The *angle of refraction* is the angle between the refracted ray and the normal line. The incident and refracted rays are defined in terms of the direction the light is going as it crosses the surface between two materials. Going from air into glass, the incident ray is in air and the refracted ray is in glass. Going from glass into air, the incident ray is in glass, and the refracted ray in is air.

Investigation 22.2 Refraction

2 Tracing rays through the prism

Create an angle of incidence of at least 25°; trace beam going into and out of prism.

A prism is a solid piece of glass with polished surfaces. Prisms are useful for investigating how light bends when it crosses from one material into another.

1. Place the prism on a piece of graph paper as shown at left. Shine the laser so the beam comes out the opposite short side. The angle of incidence should be at least 25 degrees.

2. The beam is entering the prism from the air and passing through the prism into the air again. Using a sharp pencil and an index card, carefully trace the path of the laser beam as it enters and exits the prism.

3. Remove the laser and prism from the paper.

4. Draw the lines connecting the beam through the glass as shown at right. Now, identify the incident/refracted pair of rays involved when the beam passes from the air to the glass. Next, identify the incident/refracted pair of rays involved when the beam passes from the glass to air.

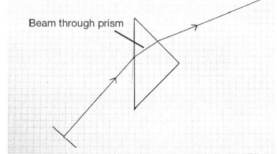

Connect the beam that travels through the prism (shown here in green)

3 The index of refraction

The *index of refraction* is a property of a material that describes its ability to bend light rays. Air has an index of refraction of 1.0. The higher the index of refraction, the more the material bends light.

a. Draw the normals to the two faces of the prism that the beam passed through as shown at right.

b. When light goes from a low-index (air) to a higher-index material (glass), does it bend toward the normal or away from the normal?

c. When light goes from a high-index (glass) to a lower-index material (air), does it bend toward the normal or away from the normal?

d. Use the two normals and a protractor to determine the angles of incidence and refraction for both surfaces crossed by the light beam. Use Table 1 to record the angles.

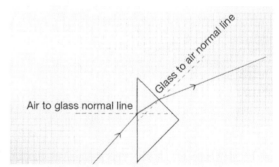

Draw the normal lines perpendicular to each surface that the beam enters and exits.

Table 1: Angles of incidence and refraction

	Angle/incidence	Angle/refraction
Going from air to glass		
Going from glass to air		

4 Another example of refraction

If you shine a laser through a cup of water with a drop of milk added, the beam will be visible. In this mini-experiment, you will see what happens when you shine the laser from air to water and back to air.

1. Fill a clear plastic cup about halfway with water. Add a drop of milk to the water.
2. Set the cup on the laminated graph sheet and trace around the base of the cup.
3. Shine the laser through the cup so it passes off-center, as shown in the photo. Use an index card and water soluble marker to find and mark the beam going into and out of the cup.
4. Remove the cup and laser. Connect the beam that passes through the cup.

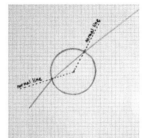

  a. Draw the normal lines to the surface of the cup at the points where the light ray enters and exits the cup.

  b. When the light is going from air into water, does the ray bend away from the normal or toward the normal?

  c. When the light is going from water back into air, does the ray bend away from the normal or toward the normal?

  d. Based on your answers to the previous questions and what you found out about the index of refraction in Part 3 of this investigation (when you used air and the glass prism), which has a higher index of refraction, air or water?

Investigation 22.3 Images from Mirrors and Lenses

22.3 Images from Mirrors and Lenses

How do mirrors and lenses form images?

A lens uses the refraction of light to bend light rays to form images. A mirror also forms an image, using reflected light instead of refracted light.

In this investigation, you will:

- use the laser flashlight to trace light rays from a lens to determine its focal length.
- show how ray diagrams are used to predict the focal length of a lens.
- use the thin lens formula to predict the locations of projected images.

Materials List

From Optics kit:
- Laminated graph sheet
- Light blue lens
- Dark blue lens
- Mirror
- Laser flashlight
- LED flashlight
- Clear filter with letter F

Other:
- Water-soluble marker

Avoid shining a laser directly into the eye.

1 The image in a mirror

When you see the tip of an arrow drawn on paper, your eye is collecting all the light rays coming from the tip of the arrow. Since light travels in straight lines, your brain "sees" the tip of the arrow at the point where all the light rays seem to come from.

The image in a mirror is an example of a *virtual* image. When you observe the tip of the arrow in a mirror, light rays are reflected. The reflected rays *appear* to come from somewhere behind the mirror. You see a virtual image of the arrow reflected in the mirror because your brain "sees" the arrow where the light rays *appear* to come from instead of where they actually come from.

1. Draw a line with a water-soluble marker on the laminated graph sheet where you will place the mirror. Place the reflecting surface of the mirror along this line. Draw a one-centimeter-long arrow on the graph paper about three centimeters away from your line. The arrow should be parallel to the line.

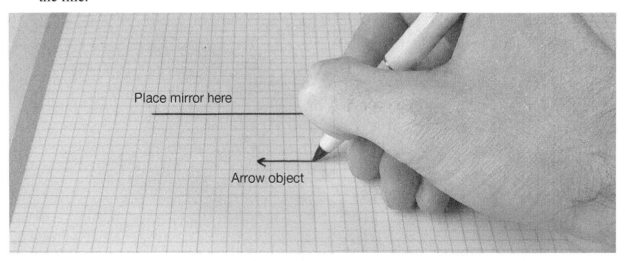

2. Move your head until you can see the reflection of the arrow in the mirror. The image of the arrow appears to be behind the mirror.

3. Hold your marker straight up with the point on the tip of your arrow. Use the marker to set the laser beam so it passes right over the tip of your arrow and hits the mirror. Trace the laser beam using the index card technique you learned in Investigation 22.1. Trace the incident and reflected beams.

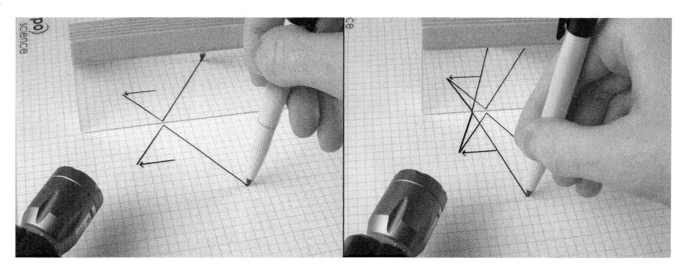

4. Move the laser so the beam passes over the tip of your arrow from a different angle, but still hits the mirror. Trace this second beam as you did in step 3.

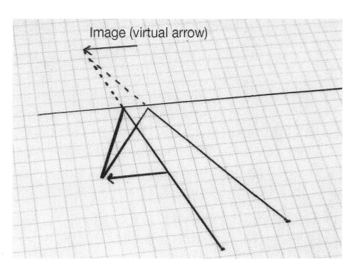

5. Remove the mirror and use a ruler to extend the two reflected rays. They should meet in a point on the other side of the line from where the real arrow is. The meeting point for the reflected rays is where you see the image of the tip of the arrow. The image forms where all rays that leave the same point on an object come together again.

Safety Tip: *Never* **look directly into a laser beam. Some lasers can cause permanent damage to your eyes.**

2 Arguing from evidence

a. Explain why the image formed by a mirror appears to come from the place where the reflected rays meet. In your answer, refer to the fact that each point on an object is the source of many rays of light. You might want to include a sketch.

b. What can the distance from the surface of the mirror to the object tell you about the location of the object's image in the mirror?

c. If you performed the same activity over the back end of the arrow instead of the tip, where do you think the two extended rays would meet? Make a prediction and try it. Was your prediction correct?

Investigation 22.3 Images from Mirrors and Lenses

3 Refracting light through a lens

Like a prism, a converging lens bends light. Because the shape of a lens is curved, rays striking different places along the lens bend different amounts. The laser allows us to follow the path of the incident and refracted rays. Rays that approach a lens *parallel to the axis* meet at a point called the *focal point*. The distance between the center of the lens and the focal point is called the *focal length*.

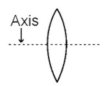

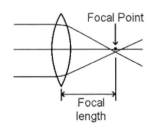

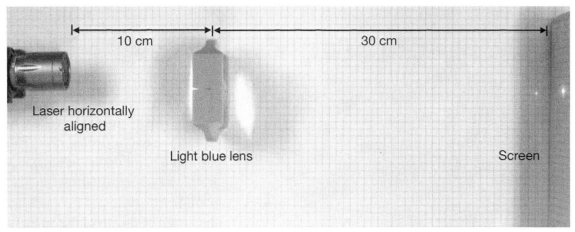

1. Divide your laminated graph sheet in half horizontally with a line. You will use the top half to experiment with the light blue lens, and the bottom half will be reserved for the dark blue lens.
2. Place the laser on the edge of the laminated graph sheet and shine the laser so it follows a horizontal grid line across the paper.
3. Place the light blue lens, with the slot facing up, 10 centimeters to the right of the laser. Line the lens up vertically using the grid lines on the graph paper. It is important that the beam is perpendicular to the lens. Make sure the beam of the laser is lined up with the middle of the lens. There are lines on the side of each lens indicating the middle of the lens.
4. Trace around the base of the lens so it can be removed and put back in place in case you need to move it to complete ray tracing.
5. Shine the laser through the lens so the beam passes off-center, almost at the very outer edge of the lens.
6. Trace the incident and refracted rays. Be sure to carefully mark the points where the beam exits the laser, enters the lens, exits the lens, and then hits the block. Connect all these points to see the path of the beam.
7. Realign the laser with a different horizontal grid line parallel to the original beam and closer to the center of the lens. Again, trace the path of the beam before and after it passes through the lens.

Images from Mirrors and Lenses Investigation 22.3

8. Realign the beam so it passes directly through the center and trace the beam again.
9. Trace two more beams passing through the lens on the other side of the center of the lens for a total of five beams. Label the beams 1 through 5 on both sides of the lens.
10. Repeat steps 2 through 9 with the dark blue lens using the bottom half of your graph paper.

4 Constructing explanations

a. Feel the glass surface with your fingers and note the shapes of the lenses. How are they different?
b. Draw a quick sketch of the shape of each lens itself with no stand from a side view. Label each lens.
c. Describe the paths of the rays before and after they traveled through the light blue lens. Include the words *refract*, *converge*, and *diverge* in your description.
d. What is the focal point of a lens? Mark the focal point on the light blue lens ray diagram.
e. The beams may not all meet at the same exact point. Find the approximate middle of where all the beams meet. Measure the focal length of the light blue lens. This is the approximate focal length of your lens.
f. Describe the paths of the rays before and after they traveled through the dark blue lens. Include the words *refract*, *converge*, and *diverge* in your description.
g. One lens is referred to as a *diverging lens*, and the other a *converging lens*. Use your ray tracing diagrams to determine which lens is a diverging lens and which is a converging lens. Research the terms *convex* and *concave*, and explain which lens is convex and which lens is concave.

5 Projecting an image with a lens

You can think about a lens as collecting a cone of light from each point on an object. In a perfect lens, all the light in the cone is bent so it comes together at a point again to make the image. This is how movie projectors take an image on film and project it onto a screen.

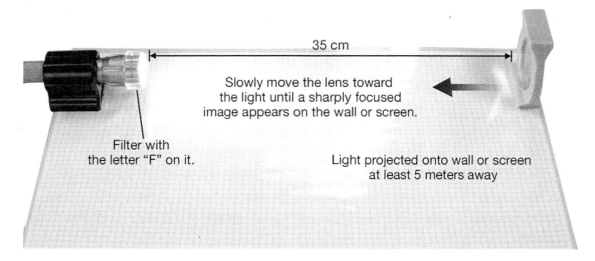

1. Place one of the lights near the edge of the graph paper. Place the clear filter with the letter *F* onto the flashlight. Turn it on and shine it horizontally across the laminated graph paper.
2. Take the light blue lens and set it on the graph paper 35 centimeters away from the light.

Investigation 22.3 Images from Mirrors and Lenses

3. Shine the light at a distant, light-colored wall at least 5 meters away. If one is not available, affix a piece of paper to the wall as your projection screen. Slowly move the lens toward the light until you see a sharp image of the *F* on the wall or screen. Have one group member check the projected image closely while the lens is slowly moved to determine exactly where the lens needs to be to make it come into focus.

a. At what distance from the light does the lens produce a sharply focused image?

6 The thin-lens formula

There are mathematical ways to predict the location of the image formed by a lens from the light reflected off of a particular object. The thin-lens equation provides a way to calculate where images form given the positions and focal lengths of all objects and/or lenses in the system. The thin-lens equation is a good approximation as long as the object and image distances are much greater than the thickness of the lens.

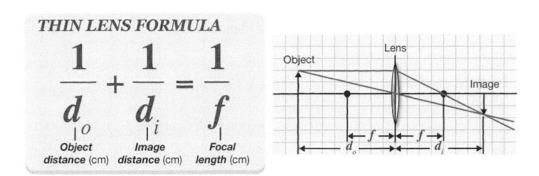

THIN LENS FORMULA

$$\frac{1}{d_o} + \frac{1}{d_i} = \frac{1}{f}$$

Object distance (cm) Image distance (cm) Focal length (cm)

When using the thin-lens equation, distances are either positive or negative depending on a *sign convention*. The equation is written assuming that light goes from left to right. When the object and image appear like the diagram above, all distances are positive.

1. Object distances are positive to the left of the lens and negative to the right of the lens.
2. Image distances are positive to the right of the lens and negative to the left of the lens.
3. Negative image distances (or object distances) mean virtual images (or objects).

The image from one lens becomes the object for the next lens. In this manner, the thin-lens equation can also be used to analyze multiple-lens systems.

Look at the thin lens formula and your answer to 4a when answering the following questions.

a. What was your object distance?
b. What was your image distance?
c. Why was the object distance in this case a good approximation of the focal length of the light blue lens? (*Hint*: What happens to a term in an equation when the denominator is very large, like the image distance here?)
d. Compare the focal length value you measured in Part 2 to the focal length value you measured in Part 4. Which one do you think is a better approximation?

7 Predicting the image distance using the thin lens formula

Use the thin-lens formula and the light blue lens to predict the image distance for four different known object distances.

1. For each of the object distances listed in Table 1, use the focal length you measured in Part 4 for the light blue lens, and plug the values into the thin-lens equation to predict the distance at which an image should form.
2. Place the screen at the predicted image distances and locate the image. How close were your predictions? Record the measured image distance for each trial in Table 1.

Table 1: Using the thin-lens equation to predict image distances

Object distance (cm)	Focal length (cm)	Predicted image distance (cm)	Measured image distance (cm)
30			
40			
50			
60			

a. How close did your prediction of the image come to the actual image? Answer with a percentage.

b. Use your actual measured image and object distances to calculate the focal length of your lens using the thin lens formula. How close was your approximation? How about the approximation from Part 2?

Investigation 23.1 The Electromagnetic Spectrum

23.1 The Electromagnetic Spectrum

What is the electromagnetic spectrum?

The *electromagnetic spectrum* is a group of waves that all travel at the same speed (3×10^8 m/s), but have different wavelengths and frequencies. The waves in this spectrum are called *electromagnetic waves*.

In this investigation, you will:

- research an important technology and the type of electromagnetic wave it uses.
- prepare a poster containing the information you discover.
- do an oral presentation to share the information with your class.

Materials List

- Poster board
- Access to the internet and other research materials
- Poster-making supplies

1 Researching your electromagnetic wave

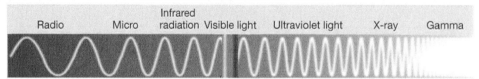

Your teacher will assign your group one industrial or medical technology that uses a specific type of electromagnetic wave that you will research using the internet or other reference materials. You will construct a poster displaying the information you gather about the technology and the type of wave it uses. You will also present what you have learned to your classmates. All the posters made by the class will be displayed in the order of the waves as they appear in the electromagnetic spectrum.

Your research should include the answers to the following questions. Do not limit yourself to these questions; include other interesting information as well. Record the sources you use for a bibliography.

1. Who invented this technology? Where and when was it invented?
2. Who discovered the type of electromagnetic waves this technology uses? Where and when was it discovered?
3. What is the range of wavelengths and frequencies for your type of wave?
4. What is the source of the wave?
5. Are these waves easily blocked, or can they pass through objects?
6. Do the waves have an effect on people? Are they harmful?
7. Discuss other uses for the waves. These may include inventions that we use in our everyday lives, medical and industrial uses, or ways scientists use the waves for research.

2 Communicating your results

a. Organize your information on a poster. You should include drawings where they are appropriate. Make a bibliography of your sources and put it on the back of your poster.

b. Present what you learned to your class. You should be able to discuss the uses for your type of wave.

c. Listen to all of the presentations. Which type of wave do you think is the most useful? Why?

23.2 The Wave Nature of Light—Polarization

What are some ways light behaves like a wave?

Polarization is a property of transverse waves. Sound is a longitudinal wave, while light is a transverse wave. To get a sense of polarization, you are going to look at waves on a spring.

In this investigation, you will:

- use a spring to demonstrate the behavior of a light wave.
- explain the interaction of light and polarizers using the wave theory of light.

Materials List

From Optics kit:
- 2 Polarizers

Miscellaneous:
- Snaky spring
- Permanent marker
- Paper or foam cup
- Meter stick

1 Polarization of a transverse spring wave

With a transverse wave, the oscillation can be in different directions relative to the direction the wave moves. Waves on a spring are a good example of *polarization*.

1. Find a partner. Each of you should take one end of the spring and stretch it. Don't let go, or the spring will snap back suddenly. *Be very careful with the springs.*
2. One person should hold the spring firmly without moving.
3. The other person should shake the spring up and down at a frequency to get the second harmonic wave. It has a shape like the picture below. This wave is vertically polarized since it oscillates up and down in the vertical direction.
4. Stop shaking the spring and let it settle down.
5. Shake the spring side-to-side at the same frequency and you will see the second harmonic wave. This time the wave is horizontally polarized since the oscillations are back and forth in the horizontal direction.

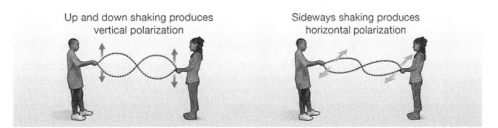

2 Arguing from evidence

a. Describe the motion of the spring using the terms *horizontal polarization* and *vertical polarization*. Your description can be in words, diagrams, or both.

Investigation 23.2 The Wave Nature of Light—Polarization

b. Suppose you try to sandwich your spring wave between two boards. What happens to the waves if you make them pass through the narrow space between the boards? If the boards were oriented like the picture below, discuss how the two different polarizations of waves would behave. Which would get through the slot and which would be blocked?

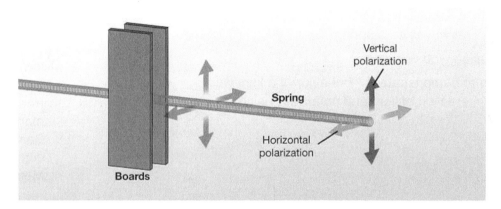

c. Describe the polarization of water waves. Are there two polarizations (like the spring) or only one? What is it about a water surface that makes it different from a spring?

3 The polarization of light

If you were to observe polarization, it would be strong evidence that light is a transverse wave. A *polarizer* is a material that allows one polarization of light to pass through but blocks the other polarization. A polarizer works like the two boards, except the "boards" are long, thin molecules. The molecules let light waves oscillate one way but absorb light that oscillates the other way.

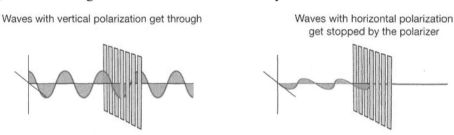

1. Take one sheet of polarizer and look through it. Observe the effect of looking through the polarizer. Try rotating the polarizer and see if it makes a difference.
2. Take a second sheet of polarizer and look through it. Observe the effects, just as with the first sheet.
3. Look through both sheets of polarizer together. Leave one fixed and rotate the other one as shown in the photo below. Observe how much light you see through both polarizers as you rotate the second one.

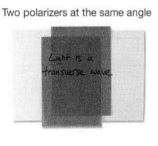
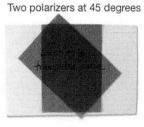
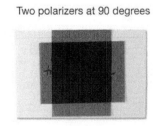

4 Constructing explanations

a. The light from the Sun (or a lamp) is not polarized, meaning it is a mixture of light that is polarized equally in all directions. Explain why the light is reduced passing through one polarizer.

b. When the light passes through the first polarizer, it becomes polarized. We say light is polarized when it consists of only one polarization. Explain why rotating the second polarizer changes the amount of light you see coming through.

5 Why polarizers make good sunglasses

1. Take a cup and draw some letters on the inside, about 2 centimeters below the rim. Draw the letters with a waterproof permanent marker. Fill the cup with water.
2. Place place the cup on a table (or the floor) a couple meters from a window. When you stand over the cup, you can see the letters through the surface of the water.
3. Position yourself so you can see light from the window reflected off the surface of the water. The glare on the surface should prevent you from seeing the letters.
4. Look at the cup (and the glare) through a polarizer as you slowly rotate the polarizer. Notice you can see the letters sometimes, and sometimes you can't.

a. Is there an orientation of the polarizer that allows you to see through the glare? If so, what is it?
b. What does this experiment tell you about the light that is reflected from the surface of the water compared to other light you see?
c. Explain how polarizing sunglasses can stop most of the glare but still allow half the regular (unpolarized) light to come through.

Investigation 23.3 The Particle Nature of Light—Phosphorescence

23.3 The Particle Nature of Light—Phosphorescence

How does light fit into the atomic theory of matter?

Can you name several things that create light? The Sun, burning logs, glowing neon gas in a sign, and light bulbs are common examples. What do all these sources of light have in common? They are all made of atoms. How do atoms create light? Work through this investigation to find out.

In this investigation, you will:

- experiment with a photoluminescent material.
- explore the quantum theory of light.

Materials List

From Optics kit:
- Laminated graph sheet
- 3 Flashlights
- Red, green, and blue filter caps

1 How is light produced?

Almost all the light you see is produced by atoms. When atoms absorb energy, electrons rise to higher energy levels. When the electrons fall back to their lower energy state, they may release energy in the form of light.

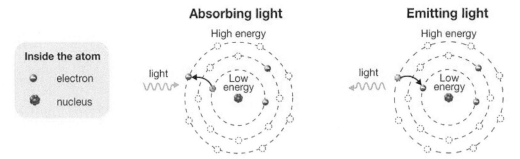

In some elements, it takes time for the energized electrons to fall back and give up their energy. These elements store energy and give off light slowly over a period of time. This is how some kinds of glow-in-the-dark materials work. Embedded in the material are chemical substances called *phosphor*. When light energy hits the atoms of the phosphor, electrons absorb energy. When the electrons fall back, they slowly release the stored energy, and the material glows until all the electrons have returned to the lowest energy level.

The Particle Nature of Light— Phosphorescence Investigation 23.3

2 Examine the effects of light on glow-in-the-dark material

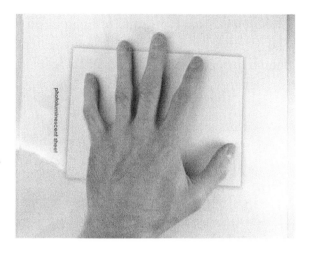

1. Place the laminated graph sheet on your table, graph side facing up. With all the room lights on, turn the sheet to the other side and quickly cover part of the material you find on the back with your hand.
2. Keep your hand in one place on the material for a minute, then turn the lights off (the room needs to be quite dark) and remove your hand. Record your observations.
3. Next, expose the material to light, completely uncovered.
4. Turn off the light and wait a minute, then place your hand over part of the material.
5. Remove your hand and record your observations.

In answering these questions, think in terms of light and energy. Explain what happens to the energy in each of these situations.

a. Why didn't the material glow where your hand covered it (step 2)?

b. What happened when your hand was allowed to rest on the glow-in-the-dark material after it was already glowing (steps 4 and 5)?

3 Examining the effect of different colors of light

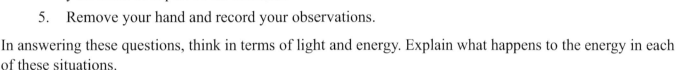

In part 2, you used a source of white light (your room lights) to add energy to the phosphorus atoms in the glow-in-the-dark material. White light is a mixture of all colors of the rainbow. In this section, you will determine what happens when more restricted colors of light are used to add energy to the atoms of the phosphor. You need to make a table to record your observations.

1. Allow the glow-in-the-dark material to stop glowing by leaving it in the dark for a few minutes.
2. Switch on the flashlight with the red cap and shine it on the glow-in-the-dark material from a distance of about 10 centimeters. Wait 15 seconds, take the flashlight away, and record your observations.
3. Try the same experiment again with the red flashlight five centimeters away and then again with the flashlight placed right on the surface. Decreasing the distance increases the intensity of the light without changing its color.
4. Repeat the procedure with the green flashlight. Record your observations.
5. Repeat the procedure with the blue flashlight. Record your observations. You will answer questions about your observations in Part 5.

185

Investigation 23.3 The Particle Nature of Light— Phosphorescence

4 The quantum theory of light

A *photon* is a small quantity of light, like a particle. You can think of a photon like a short burst of a wave. The *intensity* of light describes how many photons per second are produced (or absorbed). The *color* of each photon depends on its energy. Different colors of light are produced by photons with different energies. High-energy blue photons have shorter wavelengths than low-energy red photons.

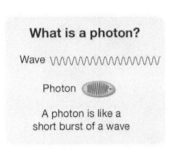
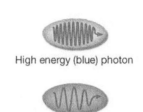
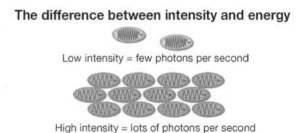

To glow, an electron in a phosphor atom must first absorb energy from a photon of light. That means a photon must have enough energy to boost the electron up one whole level to be absorbed. If the photon does not have enough energy, the phosphor atom cannot absorb it.

5 Constructing explanations

a. Based on the observations you made in Part 3, what color of light has the highest energy? Your answer should state how your observations support your conclusion.

b. You may find that the red flashlight *might* have made the photoluminescent material glow more brightly than the green flashlight. This is a discrepant (unexpected) event. Why did this happen? Propose an explanation. If time allows, you can look at the different light sources through the diffraction grating glasses that are in your kit. Observing the colored lights through these glasses will give you clues as to why the red flashlight might have made the material glow more brightly than you expected.

c. Intuitively, you might think the more intense the light is that you shine, the more brightly the phosphorus should glow. Explain how your observations support or refute this hypothesis.

d. How does what you observed support the quantum theory of light and atoms? (*Hint*: What would have happened if electrons were free to absorb *any* energy rather than just *certain* energies?)

Engineering Design Log

Why Keep an Engineering Design Log?

People who are trying to meet a need or solve a problem often follow an approach called the engineering cycle. The engineering cycle divides the design process into a series of stages that can be repeated as needed. Engineers often use a design log to keep track of their work. As you develop your own designs, keep a log to describe your ideas, gather information, ask questions, sketch plans, record results, and share findings. Your log should include the following stages:

1. Identify a Need
What problem are you trying to solve? What process are you trying to improve?

2. Research
Part of research is identifying what you know about a topic or closely related topics that may help you to design your solution. Then you may think about any additional information you need to gather and where you can find that information. What do you know about this topic? What additional information do you need?

3. Specify Requirements
The requirements of a design process include criteria, constraints, and trade-offs. Criteria are the standards a product or process must meet in order to be successful. Constraints are rules that limit the design of your product and include cost, availability of materials, and time. A trade-off is a compromise that exchanges one idea with another that may not be as good but still achieves the design requirements. For this design challenge, what are the design criteria and constraints? Can you identify any trade-offs at this stage? (Trade-offs often become more apparent during subsequent stages of the design process.)

4. Brainstorm Solutions
Describe the different ways you could solve the problem. Which solution should you explore first? Can you make a plan or blueprint of your solution?

5. Prototype
Now that you have chosen a solution, build your prototype. At this stage, you may need to gather some materials and/or tools. Did you learn something new as you constructed your prototype?

6. Test and Refine
How will you test your design and what data will you record? Once you have your prototype and other materials set up and you are ready to record your observations, conduct your tests. Construct a data table to record and analyze your data. Considering the results of your tests, what improvements should you make to your design?

7. Communicate
Scientists and engineers share their ideas within communities so that others may benefit from their work and review their conclusions. How will you share your design with others? What suggestions do others have for your design?

8. Reflect and Revise
The design process does not have to end when the results are communicated. What new information could you find that would help your design? Is there another type of test you could perform on your design, or an idea you weren't able to investigate in the first round of the process? Review your notes and reflect on these questions.

Engineering Design Log

The engineering cycle can also be thought of as a series of decisions. This diagram shows where these decisions are often made.

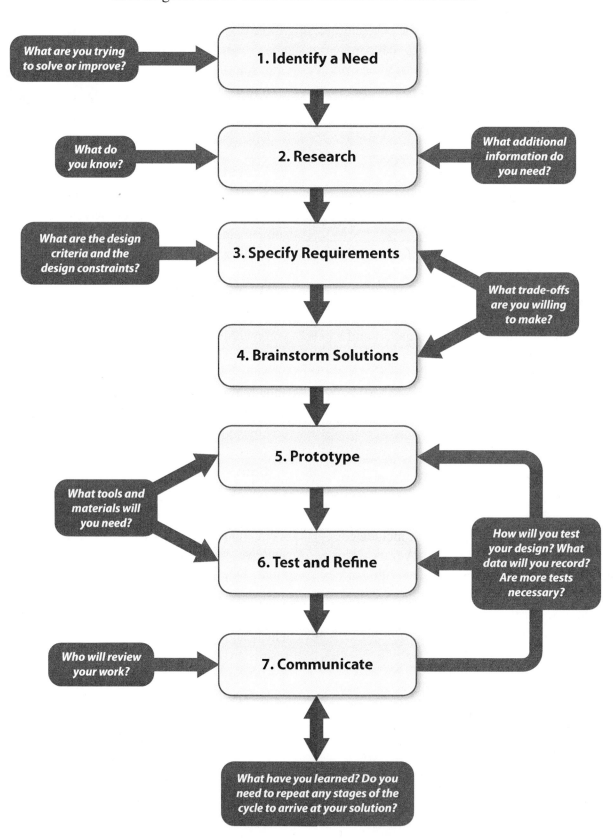

Engineering Design Log

Name _____ Date _____ Investigation [ ]

1. Identify a Need

What problem are you trying to solve?
What process are you trying to improve?

Engineering Design Log

Name _____ Date _____ Investigation ☐

2. Research

Part of research is identifying what you know about a topic or closely related topics that may help you to design your solution. Then you may think about any additional information you need to gather and where you can find that information. What do you know about this topic? What additional information do you need?

_____
_____
_____
_____
_____
_____
_____
_____
_____
_____
_____

Engineering Design Log

3. Specify Requirements

The requirements of a design process include criteria, constraints, and trade-offs. Criteria are the standards a product or process must meet in order to be successful. Constraints are rules that limit the design of your product and include cost, availability of materials, and time. A trade-off is a compromise that exchanges one idea with another that may not be as good but still achieves the design requirements. For this design challenge, what are the design criteria and constraints? Can you identify any trade-offs at this stage? (Trade-offs often become more apparent during subsequent stages of the design process.)

Engineering Design Log — Name _____ Date _____ Investigation ☐

4. Brainstorm Solutions

Describe the different ways you could solve the problem. Which solution should you explore first? Can you make a plan or blueprint of your solution?

_____
_____
_____
_____
_____
_____
_____
_____
_____
_____
_____

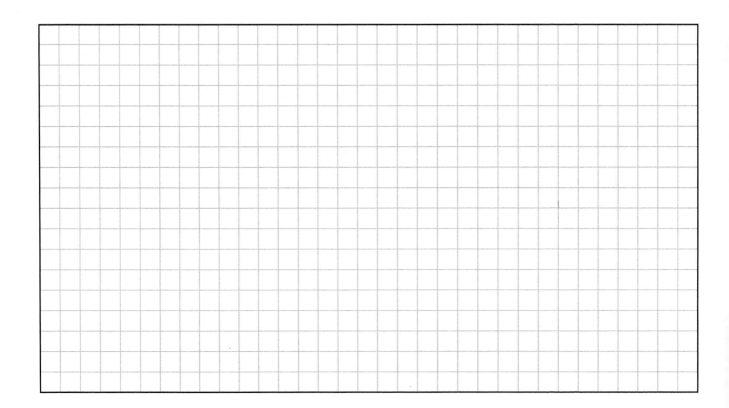

Engineering Design Log

Name _____ Date _____ Investigation [ ]

5. Prototype

Now that you have chosen a solution, build your prototype. At this stage, you may need to gather some materials and/or tools. Did you learn something new as you constructed your prototype?

_____

_____

_____

_____

_____

_____

_____

_____

_____

_____

Engineering Design Log

Name _____ Date _____ Investigation ▢

6. Test and Refine

How will you test your design and what data will you record? Once you have your prototype and other materials set up and you are ready to record your observations, conduct your tests. Construct a data table to record and analyze your data. Considering the results of your tests, what improvements should you make to your design?

_____
_____
_____
_____
_____
_____
_____
_____
_____

Engineering Design Log

Name _____ **Date** _____ **Investigation** [ ]

7. Communicate

Scientists and engineers share their ideas within communities so that others may benefit from their work and review their conclusions. How will you share your design with others? What suggestions do others have for your design?

_____

_____

_____

_____

_____

_____

_____

_____

_____

_____

_____

Engineering Design Log

Name _____ Date _____ Investigation ☐

8. Reflect and Revise

The design process does not have to end when the results are communicated. What new information could you find that would help your design? Is there another type of test you could perform on your design, or an idea you weren't able to investigate in the first round of the process? Review your notes and reflect on these questions.

Engineering Design Log

Name _____ Date _____ Investigation [ ]

Engineering Design Log Sample Grading Rubric

Stages of the Engineering Cycle	Exceeds expectations	Meets expectations	Does not meet expectations
Research and Specify Requirements	• Problem statement identifies needs • Statement of design requirements includes **at least 3** constraints and **at least 3** criteria	• Problem statement identifies needs • Statement of design requirements includes some constraints and criteria	• The problem statement is incomplete or the statement of design requirements does not include criteria and/or constraints
Brainstorm and Prototype	• Includes a diagram or plan of the prototype that indicates each of the design elements and their purpose or function	• Includes design elements indicated by a diagram or plan	• Does not include design elements indicated by a diagram or plan
Test and Refine	• The data table indicates which design was tested (ex. "1, 2, etc.") and the performance of each prototype • The data table lists trade-offs in each design • Accompanying the data table are diagrams of each design. Each diagram labels refinements made on the pervious design.	• The data table indicates which design was tested (ex. "1, 2, etc.") and the performance of each prototype • The data table lists trade-offs in each prototype	• The data table does not indicate which design was tested or the performance of that design • The data table does not list the trade-offs in each prototype • OR, a data table is not included

Chapter 3 Extension Investigation

Name _____ Date _____

Newton's Second Law

What is force? What is the relationship between force and motion?

Newton's second law is probably the most widely used relationship in all of physics. It tells us how much force is needed to cause an object's speed to increase or decrease at a given rate. It also allows us to figure out how much force is involved by watching the motion of an object change. In this investigation, you will explore Newton's second law by measuring the change in speed as a function of force and mass.

Materials:
- ✔ Energy Car and Track
- ✔ DataCollector and 2 photogates
- ✔ Mass balance
- ✔ Physics Stand
- ✔ Clay

❶ Looking at the motion along the track

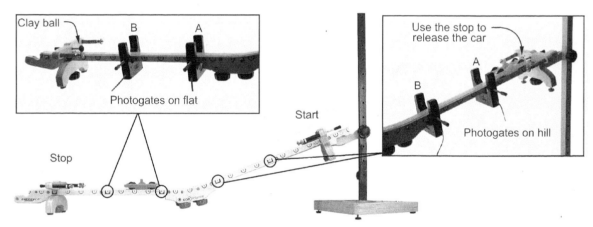

1. Set up the track with a hill and a level section.
2. Put two photogates 20 cm apart on the level section. Adjust the level of the track until the car has constant speed on the level part (same time through Photogates A and B). Roll the car and record the three photogate times in table 1. Use the stop so you can repeat the same release position.
3. Put the two photogates 20 cm apart on the downhill section of the track. Roll the car starting from the same place and record the three photogate times (Table 1).
4. Calculate the speed of the car at Photogates A and B.

Table 1: Speed data

	Time Photogate A (s)	Time Photogate B (s)	Time from A to B (s)	Speed at A (m/s)	Speed at B (m/s)
Level section					
Downhill section					

198 Newton's Second Law

Chapter 3 Extension Investigation

Name _____ Date _____

❷ Analyzing your data

a. Where is there a net force acting on the car? How do you know?

_____

_____

b. Where is there zero net force on the car? How do you know?

_____

_____

c. Can you have constant speed with zero net force? What experimental data support your answer?

_____

_____

_____

d. Write down a formula for the **acceleration** of the car in terms of the speeds at photogates A and B and the time from A to B.

e. Calculate the acceleration of the car in m/s² on both the level section and the downhill section.

f. Explain the difference in acceleration between the level and downhill sections using Newton's second law and the concept of force.

_____

_____

_____

_____

Newton's Second Law **199**

Chapter 3 Extension Investigation

Name _____ Date _____

❸ Speed and time graphs

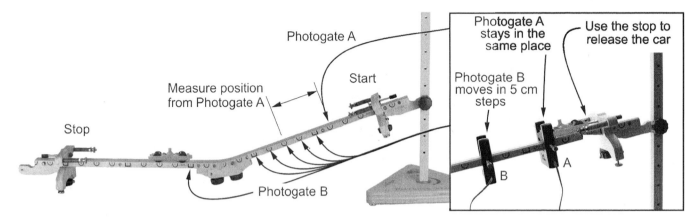

1. Set Photogate A near the top of the hill and leave it there.
2. Move Photogate B along the track in 5 cm steps and record its position relative to Photogate A. Measure the three photogate times for each position of Photogate B.

Table 2: Speed vs. time data

Position (cm)	Time through A (s)	Time through B (s)	Time from A to B (s)	Speed at A (m/s)	Speed at B (m/s)

3. Measure the mass of the car.

_____

Chapter 3 Extension Investigation

Name _____ Date _____

4. Thinking about what you observed

a. Draw the graph of speed vs. time for the track. For the *x*-axis use the time from Photogate A to B.

Title: _____

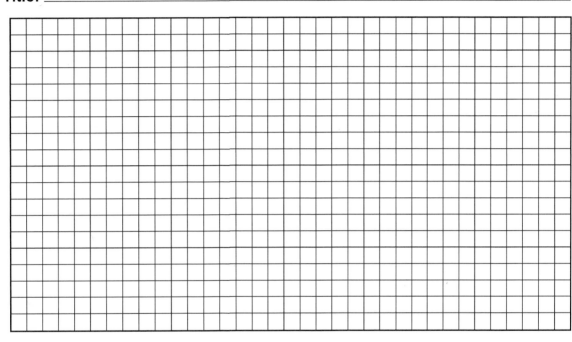

y-label: _____

x-label: _____

b. Use Newton's second law to calculate the force acting on the car at each position. Measure the acceleration from your speed vs. time graph. Where does the force come from?

Position							
Force							

Newton's Second Law   **201**

Chapter 3 Extension Investigation

Name _____ Date _____

c. Draw the graph of force vs. time and compare this graph to the speed vs. time graph. What relationship is there between the two graphs?

_____

Title: _____

y-label: _____

x-label: _____

d. Explain how speed and acceleration are different using your speed vs. time graph as an example.

_____

_____

202 Newton's Second Law

Chapter 3 Extension Investigation Assessment

Name _____ Date _____

Indicate whether the statement is true or false. If false, change the identified word or phrase to make the statement true.

1. A steeper line on position vs. time graph means a *slower* speed. _____

2. The rate of change in the velocity of an object is called *acceleration*. _____

Use this graph for the following questions. The graph shows the motion of the Energy Car in two separate trials of an investigation. In each trial, the car was launched from one end of a level track using the same number of rubber bands.

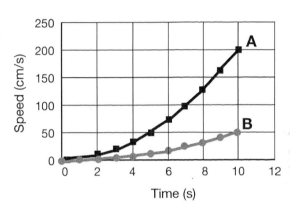

3. What is the final speed in trial B, after 10 s?

_____

4. According to the graph, how is the car moving?

   a. At a constant speed
   b. With a positive acceleration, speeding up
   c. With a negative acceleration, slowing down
   d. The car is not moving

5. The car was subjected to the same force in each trial. According to Newton's second law, what must have changed between trial A and trial B?

_____

Newton's Second Law   **203**

Chapter 4 Extension Investigation

Name _____ Date _____

Momentum and the Third Law

What makes moving objects keep going at the same speed in the same direction?

When you throw a ball it goes in the direction you threw it and does not suddenly turn one way or another unless a big force is applied. If you did try to deflect the ball you would find that for every force you apply to the ball, the ball exerts an equal and opposite force against your hand. This investigation is about **momentum** and **Newton's third law**: the law of action and reaction.

Materials:
- ✔ Energy Car and Track (2 cars needed)
- ✔ DataCollector and 2 photogates
- ✔ Mass balance
- ✔ Rubber bands
- ✔ Clay

❶ Making a collision

1. Set up the long straight track with a rubber band launcher on one stop and a clay ball on the other. Use the photogates to adjust the track so it is level (same time through A and B).
2. Put a photogate on either side of the center. Photogate A should be closest to the rubber band.

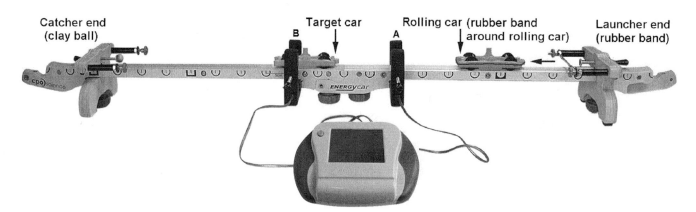

3. Place one car between the photogates. This is the target car. The pointy end of the car should be facing the launching end of the track. Place one steel ball in this car.
4. Wrap a rubber band around the second car and place it on the launching end of the track. This is the rolling car. This car should also have one steel ball and should have its pointy end facing the launcher and the "V" end facing the target car. The cars will collide with each other at the center of the track.
5. Launch the rolling car at the target car. After the collision, record the times through Photogates A and B. You may have to use the "Memory" icon to see the time if the rolling car goes through twice (bounces back). Observe the direction and motion of both cars after the collision.

Momentum and the Third Law

Chapter 4 Extension Investigation

Name _____ Date _____

❷ Thinking about what you observed

Table 3: Collision data

Mass of target car (kg)	Mass of rolling car (kg)	Rolling car before collision		Rolling car after collision		Target car after collision	
		Photogate (s)	Speed (m/s)	Photogate (s)	Speed (m/s)	Photogate (s)	Speed (m/s)

a. Consider two colliding cars of equal mass. Describe the motion of the two cars before and after the collision.

_____
_____
_____
_____

Momentum and the Third Law

Chapter 4 Extension Investigation

Name _____ Date _____

  b. The target car must exert a force on the rolling car to stop it. How strong is this force relative to the force the rolling car exerts on the target car to get it moving? What experimental evidence supports your answer?

_____

_____

_____

_____

  c. Look up Newton's third law, and state how it applies to the collision of the two cars.

_____

_____

_____

❸ Momentum

Target car
0 weights
2 weights

Four combinations

Moving car
0 weights
2 weights

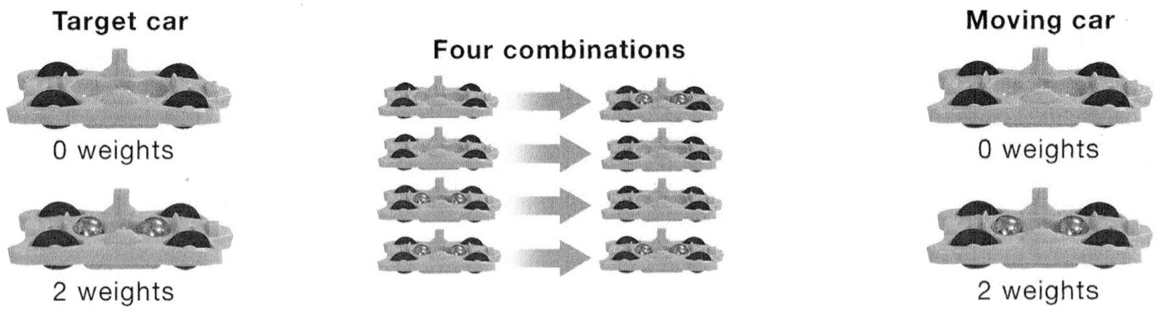

1. Try the experiment with the four combinations of mass shown above. Add the data to Table 1.
2. Try the experiment for several different speeds of the moving car. Add the data to Table 1.

❹ Exploring further

  a. Describe the motion of the two cars when the target car has more mass than the rolling car.

_____

_____

  b. Describe the motion of the two cars when the target car has less mass than the rolling car.

_____

_____

Chapter 4 Extension Investigation

Name _____ Date _____

   c. Research and write down a formula for the momentum of a moving object.

   _____

   d. Calculate the total momentum of the two cars before and after each collision. Be sure to remember that momentum can be positive or negative depending on the direction of motion.

   ✓ e. Research and write down the law of **conservation of momentum**. Describe how your data either support or do not support this law.

   _____
   _____
   _____
   _____
   _____

Momentum and the Third Law

Chapter 4 Extension Investigation Assessment

Name _____ Date _____

1. What is Newton's Third Law?

   _____

   _____

2. When you are moving on ice skates, you push against the ice with your feet. What pushes you and allows you to move forward? Use Newton's third law to explain walking.

   _____

   _____

   _____

   _____

This image shows examples A, B, C, and D. In these examples a moving car on the left collides with a motionless car on the right. Use the image below to answer the next three questions.

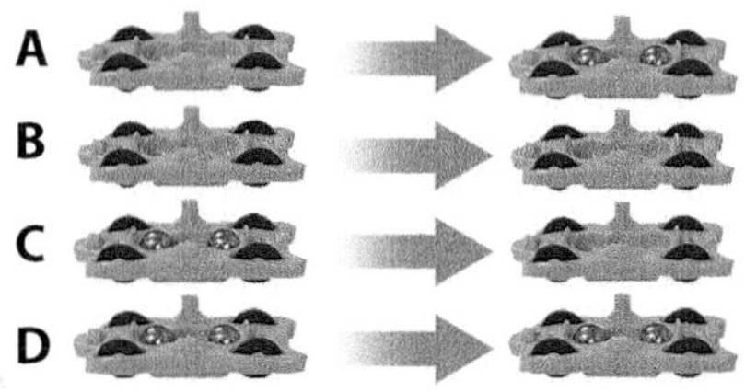

3. In situation A from the image above the car on the left has less mass than the car on the right. After the cars collide the speed of the car on the right will be _____ than that of the left-hand car.

4. In situation B from the image above the car on the left has less mass than the car on the right. How will their motion compare after the cars collide?

   a. The cars will move at equal speeds.

   b. The car on the left will move faster than the car on the right.

   c. The car on the left will move more slowly than the car on the right.

   d. Both cars will remain motionless.

5. In situation D from the image above the car on the left has the same mass as the car on the right. According to Newton's second and third laws, the acceleration of each car after the collision will be _____. Give three examples of action-reaction force pairs.

   _____

   _____

   _____

   _____

208 Momentum and the Third Law

Chapter 6 Extension Investigation

Name _____ Date _____

Engineering Design and Projectile Motion

How can we control the lateral movement of a launched marble?

In previous investigations, you determined how spring setting and launch angle affect the horizontal range of a marble flung from the Marble Launcher. In this investigation, you will apply what you have learned in a design challenge.

Materials:
- ✔ Marble Launcher
- ✔ CPO Timer and photogate
- ✔ Construction materials

- *Wear safety glasses or other eye protection when launching marbles.*
- *Launch only the plastic marbles that come with the Marble Launcher.*
- *Never launch marbles at people.*

❶ How engineers work

For this investigation, you and your group will use a process called the **engineering cycle** to design a technique for measuring and controlling the lateral, or sideways, motion of the marble during flight. You have learned that the horizontal, or forward, motion of the marble is affected by launch angle and launch speed. Lateral motion, on the other hand, is generally controlled by the long cylinder shape of the Marble Launcher. In your experimental data from Investigations B1 and B2, you probably noticed that the marble's landing site often varies along a plane perpendicular to the range of the marble (see the diagram below). This is due to the marble's lateral motion.

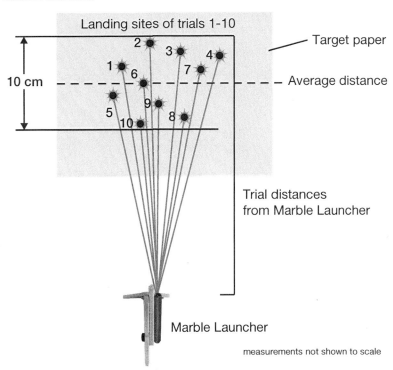

Engineering Design and Projectile Motion   **209**

Chapter 6 *Extension Investigation*

Name _____ Date _____

How could you control this variation in the marble's flight? This is the challenge of Investigation B4. **Engineers** design devices and procedures to perform tasks. This design process is accomplished using the engineering cycle. The steps in the engineering cycle are shown in this diagram:

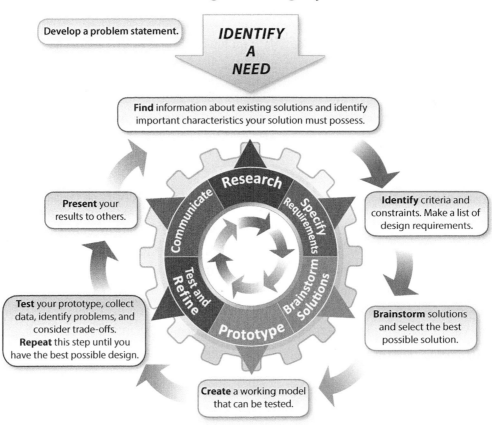

❷ **Stop and think**

✓ **a.** How might using the engineering cycle help your team design a successful process or device for controlling the lateral motion of the marble upon leaving the Marble Launcher?

_____
_____
_____
_____

Chapter 6 Extension Investigation

Name _____ Date _____

❸ Using the engineering cycle to improve your design

The goal of this investigation is to design a process or device for controlling the lateral motion of the marble upon leaving the Marble Launcher. This solution may take a wide range of forms.

Working in groups, follow the procedure below. Document the process in your notebook or Engineering Design Log as you work with your group.

1. Develop a problem statement. Your problem statement should identify the need your design will address or solve.

2. Specify the design requirements for your process. Include the **criteria**. For example, the solution you design should decrease the lateral travel of the marble after it leaves the Marble Launcher. What variables affect this travel? Can you define them precisely? **Constraints** are things that limit the design process. For instance, you do not have unlimited time in this design process, so time is a constraint. What other constraints can you identify?

3. With your group, brainstorm a process or device for controlling the lateral path of the marble's flight.

4. Describe the process or build a prototype of the device. For instance, does your process require the use of diagrams or equations? Do you need to measure angles or construct a platform to level the Marble Launcher?

5. Test and refine your design. How will you compare the function of the Marble Launcher *with* your solution to the function of the Marble Launcher *without* your solution?

6. Record any trade-offs your team had to make in the design. A **trade-off** is a compromise that exchanges one idea with another that may not be as good, but still achieves the design requirements.

7. Prepare a presentation for the class that documents your procedures and results.

Chapter 6 Extension Investigation

Name _____   Date _____

④ Extending your knowledge

✓ How did your final design compare with other designs in your class? What were some of the similarities and differences?

_____
_____
_____
_____
_____
_____
_____
_____
_____
_____
_____
_____
_____
_____
_____
_____
_____
_____
_____
_____
_____
_____

Chapter 6 *Extension Investigation* Assessment

Name _____ Date _____

1. What were the constraints of your design process?

   _____
   _____
   _____
   _____
   _____

2. What is a trade-off?

   _____
   _____
   _____
   _____

3. What was a trade-off you made or considered in your design process?

   _____
   _____
   _____
   _____
   _____

For questions 4 and 5, complete each sentence with the correct stage of the engineering cycle.

Communicate	Prototype
Research	Test and Refine
Specify Requirements	Brainstorm Solutions

4. The stage in which the timer is used is _____

5. The stage in which criteria and constraints are defined is
   _____.

Engineering Design and Projectile Motion   **213**

Chapter 10 Extension Investigation

Name _____ Date _____

Nuclear Reactions Game

How are elements organized on the periodic table?

Materials:
- ✔ Atom Building Game
- ✔ Periodic table that comes with the game
- ✔ Nuclear Reaction cards that come with the game

In this investigation, you will play a game called Nuclear Reactions using the Atom Building Game and the periodic table that comes with the game. By playing this game, you will learn about the organization of the periodic table and about two kinds of **nuclear reactions**.

❶ Introduction to Nuclear Reactions

The elements on the periodic table are arranged by atomic number, from lowest to highest. The atomic number equals the number of protons in the nucleus of an atom. The atomic number also indicates the number of electrons in an atom. Each element has a unique atomic number.

Isotopes are atoms with the same number of protons but different numbers of neutrons. Isotopes of an element have different mass numbers. The mass number of an isotope indicates how many protons and neutrons are in the nucleus of the isotope. The periodic table shows the mass numbers of the stable isotopes of each element.

Playing the game of Nuclear Reactions involves simulating nuclear reactions. To win the game, you will need to quickly figure out which nuclear reactions will make real atoms. The game is similar to the processes by which the elements on the periodic table were created inside stars. At the center of a star, nuclear reactions combine atoms to make new elements. We believe all the elements on the periodic table that are heavier than lithium were created inside stars through nuclear reactions. The process gives off a huge amount of energy that we experience as heat and light. The energy from nuclear reactions in the Sun is what makes life on Earth possible.

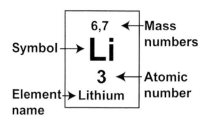

If you were to add one, two, or four more neutrons to lithium-7 you would create lithium-8, lithium-9, and lithium-11, respectively. Each of these isotopes of lithium is **radioactive**, which means that the atomic force in the nucleus (called the strong nuclear force) is not strong enough to hold these atoms together. The nuclei of these atoms fly apart. The periodic table in the kit does not show radioactive isotopes and only shows the isotopes of stable atoms.

214   Nuclear Reactions Game

Chapter 10 Extension Investigation

Name _____ Date _____

❷ Playing Nuclear Reactions

The goal of the Nuclear Reactions game is to earn points by creating neutral, stable atoms. The mass numbers of the stable isotopes for each atom are listed in the red periodic table that comes with the game. Remember that ions are atoms that have different numbers of protons and electrons—ions are electrically charged atoms.

1. Up to four players can play Nuclear Reactions. Each player should start with eight green (protons), eight yellow (electrons), and eight blue marbles (neutrons). They can be stored in the side-pockets of the Atomic Building Game board.

2. The game will last for about half an hour. The first player to earn 20 points wins.

3. To begin, each player is dealt five cards from the deck of Nuclear Reactions cards. These are held and not shown to anyone else.

4. Players take turns, choosing which card to play each turn, and adding or subtracting particles from the atom as instructed on the card. For example, if you play an "Add 2 Electrons" card, you must place two yellow marbles in the atom.

5. Subatomic particles that are added or subtracted from the atom must come from or be placed in your own pocket. You may not play a card for which you do not have the right marbles. For example, a player with only two green marbles left cannot play an "Add 3 Protons" card.

6. Each time you play a card, draw a new card from the deck so that you always have five cards. Played cards can be shuffled and reused as needed.

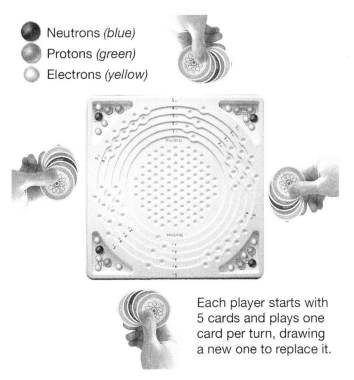

Each player starts with 5 cards and plays one card per turn, drawing a new one to replace it.

Nuclear Reactions Game **215**

Chapter 10 Extension Investigation

Name _____ Date _____

❸ Scoring points

Points are scored depending on the atom you create. You will need to use the periodic table to determine your strategy and points earned. In particular, it is useful to know which cards to play to get to stable atoms, neutral atoms, or stable and neutral atoms. The diagram below illustrates the three rules for playing the game.

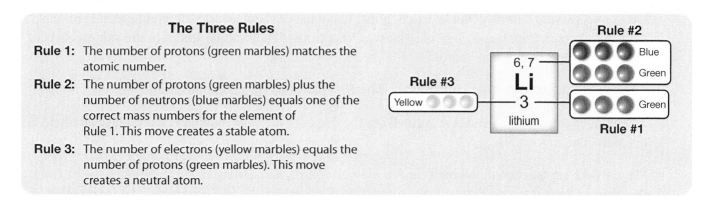

The Three Rules

Rule 1: The number of protons (green marbles) matches the atomic number.

Rule 2: The number of protons (green marbles) plus the number of neutrons (blue marbles) equals one of the correct mass numbers for the element of Rule 1. This move creates a stable atom.

Rule 3: The number of electrons (yellow marbles) equals the number of protons (green marbles). This move creates a neutral atom.

You score 1 point if your move creates or leaves a stable atom. For example, you score 1 point by adding a neutron to a nucleus with six protons and five neutrons. The resulting atom is carbon-12 which is stable. The next player can also score a point by adding another neutron, making carbon-13. To get the nucleus right you need to satisfy rules 1 and 2.

You score 1 point for adding or taking electrons or protons from the atom if your move creates or leaves a neutral atom. A neutral atom has the same number of electrons and protons. Getting the electrons and protons to balance satisfies rule 3.

You score 3 points (the best move) when you add or take particles from the atom and your move creates a stable AND neutral atom. For example, taking a neutron from a stable, neutral carbon-13 atom leaves a stable, neutral carbon-12 atom, scoring 3 points. In other words, you get 3 points if your turn makes an atom that meets all three rules.

Nuclear Reactions Game

Chapter 10 Extension Investigation

Name _____ Date _____

❹ Additional rules

Taking a turn: When it is your turn, you must either play a card and add or subtract marbles from the atom or trade in your cards for a new set of five.

Trading in cards: You may trade in all your cards at any time by forfeiting a turn. You have to trade all your cards in at once. Shuffle the deck before taking new cards.

Using the periodic table: All players should use a periodic table to play the game.

The marble bank: You may choose either of two versions of the marble bank. In version 1, players take marbles from the bank at any time so that they have enough to play the game. In version 2, players must lose a turn to draw marbles from the bank and may draw no more than five total marbles (of any colors) in one turn.

❺ Nuclear reactions

There are two kinds of nuclear reactions, **fusion** and **fission**. These kinds of reactions involve only the nuclei of atoms. (The word *nuclei* is the plural form of the word *nucleus*.) In fusion, two elements with small mass numbers combine to make an element with a larger mass number.

In the diagram above, nuclei are fused, a particle is emitted, and a lot of energy is released. The reaction in the diagram involves the fusion of hydrogen-3 (1 proton + 2 neutrons) with hydrogen-2 (1 proton + 1 neutron) to make a helium-4 nucleus, a neutron, and energy. The black dots are protons; the gray dots are neutrons.

In fission, an element with a large mass number splits into elements with smaller mass numbers. The diagram at right shows a nuclear fission chain reaction. Nuclear fission can be started when a neutron (gray dot) bombards a nucleus (black dot). A chain reaction results. A free neutron (step A) bombards a nucleus (step B), and the nucleus splits into nuclei with smaller mass numbers. More neutrons are also released (step C). These neutrons then bombard other nuclei. Nuclear reactors control fission (and energy production) by capturing neutrons to start, slow, or stop the chain reaction.

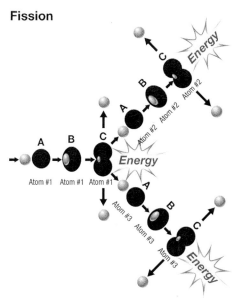

Nuclear Reactions Game   **217**

Chapter 10 Extension Investigation

Name _____ Date _____

Now use the Atom Building Game to help you work through the following questions, which will help you understand fission and fusion.

✓ a. Demonstrate the fusion reaction diagram using the Atom Building Game board. Collect enough marbles (protons, neutrons, and electrons) to build a hydrogen-3 atom (this is a radioactive isotope). Then collect enough marbles to build a hydrogen-2 atom. Place all these marbles in the correct places on the Atom Building Game board. Remove one neutron and hold it in your hand. What element is represented on the board? Why was it important to take away one neutron?

_____

_____

_____

b. Collect enough marbles (protons, neutrons, and electrons) to build lithium-6. Then collect enough marbles to build boron-11. Place all these marbles in the correct places on the Atom Building Game board. What element is represented on the board? Is this activity an example of fusion or fission?

_____

_____

c. Is the atom that results from the combination of lithium-6 and boron-11 a stable or a radioactive isotope? Is the atom an ion or neutral?

_____

d. Now build boron-10 on the Atom Building Game board. How many protons, neutrons, and electrons would you need to add to the board to make fluorine-19? If you were to add these subatomic particles to boron-10, would this represent fusion or fission?

_____

_____

_____

✓ e. Suppose you split a uranium-238 atom. If you have to break it into two pieces, name two elements that could be formed. Be sure that your two elements use up all the neutrons and protons in the uranium. Is either of your two elements a stable isotope or is one (or are both) radioactive?

_____

_____

_____

218   Nuclear Reactions Game

Chapter 10 Extension Investigation Assessment

Name _____ Date _____

1. Identify the mystery atoms that match the clues. Some clues may have more than one right answer.

   a. The atom has an atomic number of 29.
   _____

   b. The atom has a mass number of 54.
   _____

   c. The element has four isotopes.
   _____
   _____

   d. Atoms of the element each have 65 protons.
   _____

   e. Atoms of the element have three neutrons.
   _____

2. Describe the organization of the periodic table. How does this organization help you find information about elements?
   _____
   _____
   _____
   _____
   _____
   _____

3. Define the term *radioactive*. Explain why atoms of certain elements undergo radioactive decay.
   _____
   _____
   _____
   _____
   _____
   _____